Cooking in Russia

YouTube Channel Companion Reader

VOLUME 3

by
Greg Easter

A chef who doesn't understand anything about the chemistry of cooking is like a software engineer who doesn't know how a computer works.

You are never going to reach your potential without comprehending what's going on inside that "black box"... whether it's a pressure cooker or a CPU.

TABLE OF CONTENTS

INTRODUCTION TO TOPICS

As with the first volume, this was written first and foremost to provide additional information to the video recipes online at:

www.youtube.com/user/cookinginrussia

What's Different About This Volume

There are several key differences here in Volume 3 that I want to begin by spelling out.

References to Previous Volumes

Volume 2 was essentially a stand-alone book with relatively little need to be familiar with Volume 1. In this volume there are several sections that are direct continuations from Volume 2, as well as many references made to both previous volumes. This will be even more the case in the upcoming Volume 4, which will absolutely depend on you having this volume—especially for the *Spices & Coatings* recipes, some of which are not even used in this volume.

Finally...Actual Food Chemistry

In the previous two volumes I kept the technical aspects simple enough that readers didn't need to actually study anything as though it was a college textbook. We are now at an impasse. There is no way to venture deeper into these topics without you being up to speed in some of the basics of organic chemistry. Up until now I have been writing at a level that assumes the reader knows almost nothing about chemistry or physics. This can't continue if you are going to understand actual food chemistry, so part of the function of this book is to lay that necessary groundwork. If you ever wanted to understand food chemistry, this will be a milestone.

The single most challenging aspect of writing this book was finding a middle ground between being too simplistic and overwhelming the reader with what would surely seem like an ocean of technical jargon. I have erred on the side of the latter, I must confess. For those who only wanted some recipes, you can just skip the front end. For the rest of my readers, I am confident that there has never been another cookbook that explains these matters in such unflinching scientifically accurate detail. Although there are parts that most readers will find alien at first, I have tried to make these difficult concepts understandable to anyone who's interested enough to read through the explanations carefully.

A chef who doesn't understand anything about the chemistry of cooking is like a software engineer who doesn't know how a computer works. You are never going to reach your potential without comprehending what's going on inside that "black box", whether it's a pressure cooker or a CPU. Although many people have an instinctive fear of science and mathematics, this won't be *that* difficult if you are sincerely interested.

This volume primarily covers different types of flavor molecules and the most important chemical reactions in cooking. The topic of different sugars and how they react will be picked up in Volume 4.

Try not to be intimidated by the chemical drawings as you flip through these pages. When you read the accompanying text, you'll see that I have explained everything in simple language starting with the basics. Remember that you don't need to fully understand everything. There's not going to be a test! As long as you get the gist of it, you'll still be more knowledgeable than the vast majority of trained professional chefs. Seriously.

Advanced Topic Sections

There are a few places where I felt that I might be going over the head of most readers, so I divided off these technical aspects and I give you my assurance that you that you really don't need to know those details. But in case you are curious, it's there for you. Maybe you'll come back to it later. The information won't change.

One Omission

The only recipe video during this period that is not included in this book is the *Viennese Mushrooms and Cream*. The reason is that this incorporates some of the more advanced techniques that won't be covered until Volume 4.

Tertiary Additives

Unlike Volumes 1 and 2, the seasoning blends in this volume are not just a matter of grinding some spices and herbs together to produce a pleasant harmony—not that there's anything wrong with that. Spice blends are vital to gastronomy. But this is something that you've probably never heard of before, unless you are an industrial food chemist: *Tertiary additives*. Understanding this represents a quantum leap in the skills of any cook—a quantum leap that very few ever make because it is not taught in culinary academies and not something you learn by working as a chef. This is university level food chemistry, and arguably the single best thing about this book because it opens up a new world. It is the way in which commercial food manufacturers make those addictive products they profit from—only here you can engage in that same chemistry using natural food products instead of bottles of chemicals, so it's even better.

I am introducing a technique of spice blends in which flavors are absorbed onto powders. This is based on the same principle as column chromatography (see page 90). It enables you to create background layers of flavor that utilize some of the same chemicals as commercially manufactured foods, but without any of the synthetic concentrates or weird aftertastes. I realize that some of these will look bizarre at first glance, with ingredients you would never think of combining like this—but after you try these, you'll realize that this is probably the single greatest advance in cooking you will ever encounter. You will discover that you can apply these mixtures to my previous recipes (as well as your own beloved recipes) and achieve new dimensions of flavor that you never imagined could be coaxed out. Yes, it is some extra work to make these, and you may even burn some and have to start over, but when you get it right you will be stunned by the power you now wield to make previously mundane recipes sparkle with fresh life.

Some Additional Notes

Why You Don't See More Cookbooks Like This

Being a trained chef and having a doctorate in organic chemistry is a pretty rare combination, but that's not the main issue here. Very few chefs ever attempt to develop revolutionary techniques, even though they are in the ideal laboratory with the opportunity for dozens of experiments to be carried out every day. They are there to do their job, collect their pay and live to live another day. Among those few who are still sufficiently motivated after 10 to 20 years of that daily grind when they finally have the experience and status that would allow them such freedom, most have developed a set pattern of following procedures and focusing only on the tasks that a head chef has to do to keep their job. They might actually discover something new, but then they lack the scientific education to appreciate the significance of what it is they stumbled on, as in the case I described in the history of the dish on page 209 of this book.

Finally—and here is the key—among that tiny minority who both experiment and keep good notes, almost none are willing to share what they learned because of the time and work that they've invested. Giving it away feels like handing out hundred dollar bills to strangers on the street. It goes against normal human instincts. I still get twinges of that gut feeling myself when copying passages from my notes into these books, flashing back to days of blistering heat in steam-filled cramped kitchens, holding a stubby pencil in freshly burnt fingers, short on sleep for weeks as I frantically scribbled those notes down between orders while trying to tune out calls of "Chef! Chef!" in my ears. However, I'm retired now and the past is irrelevant. I am commited to passing this information along while I still can.

Three Advantages

You have three advantages here compared to other cookbooks. First, you can see what each phase of most recipes should look like in a video. If a picture is worth a thousand words, then a video is worth a million. Second, precise details on how hot the stove is set minute by minute, which is quite rare. Third, you can ask the author questions directly and receive a fast response 365 days a year!

Respect the Recipe

In a recent discussion on a message board for chefs, someone asked what the most annoying thing was about the profession? The responses were probably the last thing most people would have expected. Is it the long hours and missed birthdays and holidays? No. Is it the low pay? No. Is it the sweltering hot kitchens? No. Is it standing on your feet for hours doing repetitive tasks? No. Is it the frequent burns and cuts? Nope. Those things are all accepted as part of the job. So what is the most annoying thing? ...The lack of recognition by the general public for the skill required to do the job. People who grill some steaks on the weekend, or make family meals a couple of times a week often think of themselves as "amateur chefs". This is like someone who puts a band-aid on a child's scratch thinking that they are an amateur doctor. The field is deceptive because most people have some experience cooking, but I have yet to meet the home cook who can turn out five different meals in 15 minutes and have them all hot and perfectly presented simultaneously. Especially when that's just for one table out of many more that they are in various stages of production. That sort of organization and flawless preparation hour after hour without a mistake is every bit as unimaginable to a home cook as confidently landing a jumbo jet in heavy fog is to a 10 year old on a bicycle.

Most of the time this doesn't really bother me, though. I realize that it's human nature to not appreciate things that one hasn't actually tried to do. I had an egocentric manager at a restaurant brag that he could do a better job as a cook than any of the kitchen staff. One day we called his bluff. He was put in a uniform and handed three orders to make, much to the amusement of the cooks who stood around watching him make a fool of himself. After half an hour, even with some help, the only thing he'd made was a gigantic mess. Finally, in recognition that customers would be complaining about the long wait, he stepped back and watched us. Everything was cleaned up quickly and the orders went out like clockwork within a few more minutes. He never opened his mouth again about being able to do things better. If everyone had that experience, people would know better than to think of themselves as amateur chefs. Still, how people think of themselves is rarely an issue.

This only time such self-aggrandizing rears its ugly head is when people insist on making changes to a recipe that seriously compromises—or just plain ruins—the dish. After years of answering feedback on YouTube, I'm getting somewhat used to the sort of questions that could only be a silly joke by any restaurant cook. I've had ridiculous substitutions reported, like Bailey's Irish Cream in place of vermouth, and read questions like, "Can I leave out the red wine in Beef Bourguignon?" There are a lot of people who truly don't know. Okay. I get it now. I know that's why I'm here.

The crux of the problem is that many amateurs don't respect the recipe—either the work that went into developing it, or the skills of the chef who created it in the first place. This in itself is the mark of an amateur. When a professional is given a recipe, it is studied carefully ahead of time, mentally rehearsed and then followed to the letter. Usually a couple of times before any changes are attempted. Of course that's assuming the recipe came from a reliable chef, because then you know it's already been honed.

My advice is not to make substitutions or leave ingredients out. No, you can't use chicken stock in place of wine. If you don't want to cook with alcohol then find a recipe that doesn't call for any. If you are still asking questions like, "Can I use a crock pot instead?", or "Can I use chicken breasts rather than pork neck?", then do yourself a favor and stop trying to invent substitutions because you aren't even close to that skill level. It's not an insult any more than saying you can't land a jumbo jet in fog. Hopefully after reading this complete series you will have a better understanding, though.

Measurement Conventions

Unlike the previous two volumes, I have written out "teaspoon" instead of a lower-case "t" in most every instance to help make this more clear. Tablespoon has been left abbreviated as "T".

I will repeat this yet again:

While many cooks use terms like teaspoon and tablespoon very loosely, that is not how professional chefs operate. A teaspoon (abbreviated with a lower case t) is 5 cubic centimeters, or 5ml. A tablespoon (abbreviated as an upper case T) is 15 cubic centimeters, or 15ml, or 1/2 ounce.

GERMAN VS. JAPANESE KNIVES

This is a simple topic that has surprisingly few straightforward answers online. There are only three important differences:

Metal Strength: For knives, this is generally measured on the Rockwell scale, which is based on how much pressure it takes for a diamond to scratch or dent the metal. Japanese knives are about 10% harder. The good thing about this is that they will keep their edge longer. The bad thing is that they are more difficult to sharpen and prone to being damaged during sharpening. I suggest you use a professional knife sharpening service. You will not compete with someone who does this all day for a living.

Angle of Blade: The standard for almost all German knives is 20 degrees, while Japanese knives are typically 15 degrees, and sometimes less. A smaller angle means a sharper knife. However, if you are hacking your way through heavy joints, the edge will deform easier and will require professional sharpening sooner. For this reason most butchers stick with German knives.

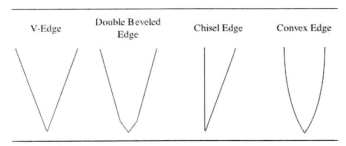

| V-Edge | Double Beveled Edge | Chisel Edge | Convex Edge |

Thickness of Blade: German knives are thicker, which gives them the advantage when pounding through tough membranes and joints. When it comes to slicing a tomato, your advantage will be with the thinner blade, so it depends on the task at hand. After you get used to working with Japanese knives, picking up a German knife will remind you of picking up an axe. But that's okay, because sometimes you need an axe. Just not when you are slicing tomatoes.

The Solution: If you are a serious cook, you need both a German and a Japanese chef's knife. If only one, choose the latter. Other basics include a heavy German boning knife plus a lighter, thinner Japanese boning knife for more delicate work, such as with fish. Most of the knife types sold are never used by professionals.

Unless you have a lot of practice, you will need a good paring knife. Seasoned professionals use a chef's knife for paring too, but that's a good way to have a serious injury if you aren't madly adept at such dexterous manipulations.

A cleaver is also essential _if_ you are going to be doing heavy butchery work. Otherwise it will sit in a drawer collecting dust.

Granton Edge: The indentations along the sides on many modern knives (as shown in the photo above) help food fall away from the blade. The type of knife shown here is a Santoku, which is a general purpose Japanese knife. Unless you are used to using this sort of knife, stick to a regular chef's knife.

Stainless Steel Knives: These lose their edge very quickly and are very difficult to sharpen. In recent years many companies have copied the visual design of high quality knives, such as the one shown in the image above here. Some organize parties the way that Tupperware used to be sold to housewives (_e.g._ The Pampered Chef). Don't fall for this. No chef or knowledgeable cook would use one of these for anything other than chasing the salesperson away. To add insult to injury, many of these knives sell for as much as a professional grade knife goes for.

THE HIGHLY VARIABLE POTATO

As humans, we have a natural tendency to want to simplify things into easy-to-remember categories and definitions, but the fewer labels you use, the more errors you are bound to have. Trying to fit all potatoes neatly into the two categories of waxy or starchy is a perfect example of the folly of oversimplification. Not only are many types in between these two, but there are other important qualities that are not reflected in those labels, including:

- Age of the potato
- Cell size (a huge range from 5 to 100 microns)
- Storage conditions (temperature, humidity, light, gases)
- Sugar content (also affected by storage conditions)
- Type of fertilizer used and mineral characteristics of the soil

Entire scientific journals that are devoted exclusively to potatoes have published research articles every month for decades in relation to genetic variations, commercial farming methods and a mind blowing array of factors that all influence the end product. In almost every article one encounters a similar caveat to the research. Here is an example from a recent article in the *European Potato Journal* trying to categorize Yukon Gold potatoes:

> *"The results suggest that Yukon Gold potatoes behave more like waxy potatoes. However these results are for only one lot of commercial potatoes, and it is well known there is significant variation in density and starch content even between potatoes of the same variety."*

When it comes to potato recipes, you can't robotically follow procedures if you want optimum results. Cooking times will vary, and the potatoes from one source can be quite different from the same type from another source, even in the same season of the same year. There is nothing you can do. Just keep it in mind.

Finally: **Always store potatoes far away from onions!** Both emit gases that cause the other to rot faster.

GROUND DRIED VEGETABLES AND MORE AS SEASONINGS

(CONTINUED FROM VOLUME 2, PAGE 24)

Dry Roasted Apples

This is another way to add deep caramelized flavors to a dish, other than brown sugar. This is especially useful in Mexican Mole sauces. You can save yourself some time by starting with sliced dried apples, but be sure they do not have any sugar or preservatives. If you are using fresh apples, core them and then slice them into wedges. You can leave the peel on. Either way you dry them on racks in the oven at 115°C (240°F) overnight, and possibly longer. They will remain leathery until you remove them from the oven and let them stand in the open air at room temperature for at an hour or more. If they are not dry and brittle enough to crack into pieces then return them to the oven. The ones you get that were already "dried" will finish drying and roasting in about 8 hours. Fresh apples can take up to twice as long.

Apple Salt

In the previous volume of this series I mentioned adding salt before roasting, and said only to do so if directly instructed. The reason is that the result is quite different. Salt draws out moisture, so the drying process is faster, but the product is different.

One example of this an ingredient I will call Apple Salt. To produce this, you slice green apples 2 mm (1/10 inch) thick. You can leave the peel on, but make sure you have removed the seeds and stems. For every 50-60 grams (2 ounces) of apple slices, grind 1 tablespoon of salt in an electric spice mill until it is a fine powder, and add it to the apples, tossing to coat evenly. Lay them on a silicone mat on a baking tray in a single layer. Roast these in an oven at 120°C (250°F) for about 2 hours with fan assist on. Then remove them from the oven and allow to stand out at room temperature for 30 minutes. Grind the pieces up in the spice mill. This ingredient is not sold commercially, but it is used in commercial food manufacturing. See *Austrian Spice Blend* (pages 220-221).

Fennel Salt

Put 120 grams (4 ounces) of the tougher outer layers of fennel and some stalks (but not the green fronds which will burn) into a food processor and grind to very small pieces. Add 1 tablespoon of table salt and grind more. Scrape out the contents to a bowl and mix more with a spatula by hand. Now spread it out on a silicone mat (Silpat) that's on a baking sheet. Put this into a 120°C (250°F) oven with fan assist ON for 45 minutes. Remove the tray and mix the material around well. Return it to the oven for another 30-40 minutes. Now take it out and let it stand at room temperature for an hour or so. Grind the mixture with 2 teaspoons of coarse salt added in an electric spice mill. Now pass it through a sieve. Discard solids that won't pass through. Yield is about 30 grams (1 ounce).

Salted Dried Limes

Slice limes across into 2.5mm (1/10th inch) thick rounds. Place on a wire rack and sprinkle with salt. Don't saturate them with salt, though—just a light sprinkling. Place in a 88°C (190°F) oven with fan assist on for about 10 hours. At this point the slices should be very dark and quite crisp, but not actually burned. If they are black like charcoal then your oven is running hot and you'll have to start over again. These slices keep a long time as long as you don't grind them up and they are protected from air, but putting them in the refrigerator will help them retain maximum flavor even longer. This is a key ingredient in several of the spice blends later on here.

Toasted Shallots or Onions

Slice shallots or white onions 3mm (1/8th inch) thick with a mandoline. Place on a wire rack in the oven at 120°C (250°F) with fan assist ON for 20 minutes. Don't go longer or they will burn. They will not be fully dehydrated, but they are best used just like this.

Dried Anchovies

Difficult to make, but available to buy. Try Russian stores, as they are a popular snack Russians have with beer. See *Sicilian Spice Blend* (page 226) and *Worcester Spice Blend* (page 227).

Dried Chipotle Chilies — See notes on page 282.

A tiny bit

~~ALL~~ ABOUT BEER

Like wine, the subject of beer is complex because it is an ancient art going back thousands of years. After decades of American beer being the thin yeasty yellow water associated with ball games and pizza joints, in recent years beer has finally started to be taken seriously again. Microbreweries have became a vibrant cottage industry and have simultaneously increased public interest in European brews, many of which have been around for centuries. Still, most people really don't know what the designations mean, or even the basic process of how beer is made.

Synopsis of the Brewing Process

The essential five steps are:

1. Soak the malted barley in warm water.
2. Add hops and boil.
3. Cool solution and add yeast.
4. Ferment.
5. Filter and bottle.

Sounds easy, doesn't it? There are big fat books devoted to the details of these operations. Master brewers spend their entire life learning the craft. The five steps you see here are like learning five words of a language.

For example, the first step calls for malted barley. What is malted barley? It means that the barley has undergone what is called malting, or sprouting. The seeds have been tricked into starting to grow. In order to do this, the seeds are soaked in water and kept warm for several days, turning them occasionally to make sure they are aerated and not rotting at the bottom. Now you have what's called *green malt* in the industry. This is transfered to large kilns to dry. The temperature will determine what kind of malted barley is produced, but it is always hot enough to stop any further germination. Sometimes it's roasted at a high temperature for making stout ales. Malted barley is sold to breweries—even most large breweries don't make it themselves. It is a specialty craft.

Malted barley is rich in sugars and starch. It also contains an active enzyme called *diastase* that was present in the seeds. By soaking the malted barley in warm water, the diastase will convert some of the starch into more sugar, which will later become food for the yeast.

Now you have a liquid that's called *wort*. In order to stop the diastase from going too far, the mixture is heated to a boil at what is judged to be the right time (depending on a great many factors). This is also when hops are added. The type of hops, and how they have been treated before they are added are also variables that the brewer must decide on.

So finally the stage is set and it's just a matter of letting the yeast do its thing, right? No. There's still the matter of what kind of yeast, the temperature it will be brewed at, and how long it will brew for. Also the type of material it will be brewed in. Although these days metal tanks are almost always used, that's quite a recent development. In the past beer was brewed in oak barrels. For lighter beers where the oak would overwhelm the brew, the barrels were lined with pitch. In more recent times, barrel aging has been put off to the end of the brewing process, and the barrels that are used previously contained bourbon, cognac, sherry and even tequila. These barrel aged versions of the brew are then sampled and blended together to produce a final product. Depending on how much care goes into this and expensive of the ingredients that were used, the final price of a bottle of beer can be more than some wines. Now you know why.

Beer vs. Ale

The key difference is the type of yeast and the temperature. Most mainstream beers use a type of yeast (*Saccharomyces Uvarum*) that sinks to the bottom of the fermentation tank and the process is carried out at a low temperature (4.5-12.7°C / 40-55°F) for a long time—often months! Bottom fermentation is the oldest type of beer and it remains the most popular in the world due to its light refreshing taste.

Top fermentation uses a different type of yeast (*Saccharomyces cervisiae*) that floats at the top. The fermentation is carried out at a slightly higher temperature (15-25°C / 59-77°F) for a shorter time (typically a week). Top fermentation produces more organic flavor molecules, especially a type called *esters* (more about those later in this book). These beers are stronger and have much more pronounced flavor. Some are quite rich and are less the type of beer you would guzzle on a hot day, and more the type you might enjoy like wine. Of course I'm speaking in generalities here. I've seen people chug potent ales like they were drinking Coors Light (which is as close to a beer *not* being beer as beer gets).

Barley Wine

This has become a popular term, but as Britain's most famous beer expert, Martyn Cornell, said, "I don't believe there is actually any such meaningful style as 'barley wine'." That is, it is just another name for a dark ale. Back in 1870, Bass Ale was marketed as barley wine, which is apparently where the name came from. The alcohol level is generally high (6-12%) and there is more raisin-like sweetness, with some even resembling port wine.

Beers and Ales for Cooking

The biggest problem with beer for cooking is that it is prone to becoming very bitter on reduction, which is inevitable if you are cooking something for very long. This is largely due to the hops in the beer. Bock, Dopplebock and Belgian ales are low in hops, so they tend to work much better in cooking. Arguably the easiest beer to cook with and obtain good results is Dopplebock. See the recipes for *Dopplebock Beef* (Volume 2, page 114) and *Dopplebock Beef Potatoes* (Volume 2, page 116).

The exception is when you actually want to introduce some bitterness as a counterpoint to other flavors, as in the *Rutabaga in Porter Ale* recipe on page 167 of this volume.

The Language of Beer

As in the case of all food and wine, most of the flavor is actually perceived in the nose. If you have ever had a severe head cold and been almost unable to taste anything, then you know this is true.

There is an extensive appendix explaining dozens of terms used in beer and wine tasting starting on page 257 of this book. On the opposite page here is a quick guide to the main types of beers and ales on the market today. A chef should be familiar with all of these.

LAGER

Pilsner is a type of Lager. This is the most popular beer and the lightest in style. It is what most people think of as "beer".

BOCK and DOPPLEBOCK

Less hops and more malt make this German style lager sweeter and stronger. Dopplebock is even stronger and sweeter than Bock, often being referred to as *liquid bread*.

HEFEWEIZEN

Wheat beer. Often bottle fermented with yeast left in the bottle and frequently flavored in a manner similar to gin. Hoegaarden is an especially popular brand that is cloudy white and lightly flavored with coriander and orange peel.

STOUT

Made from roasted barley or malt and higher than average in alcohol. Guinness is the most popular example of this style.

IPA or INDIA PALE ALE

A lot of hops for a bitter beer that's usually low in alcohol. Originally popular in India, but the style is now sold all over.

MICROBREWERY CRAFT ALES

Most of these are from America and Britain and are designed to appeal to a niche market with gimmicky hipster-friendly names and eye catching label art. Many are bitter from a lot of hops, which is an acquired taste. Some have peculiar flavors like coffee and elderflower. A few are great, but most are not.

TRAPPIST ALES

Made by monks, primarily in Belgium. These tend to be very strong (over 8% alcohol is not unusual) with little or no hops. They are often unfiltered and have spice and herbal notes.

KRIEK

This is the most common Lambic ale. It is made with sour cherries. Lambic ales are made a different way and do not taste very much like beer at all. They are sour and fruity.

HOPS

These days hops are used in nearly all beers, but that was not always the case. The first known use was around 1100AD. Before then a wide variety of bitter plants were mixed together, including dandelion and ivy. The flavor was not the most important thing, though. These plants (including hops) have natural antibacterial properties. Otherwise the beer would be overtaken by bacterial growth before it finished fermenting.

There are over twenty different popular types of hops in use by brewers today. They vary in acidity, bitterness, aroma and flavors ranging from astringent and herbal to sweet and fruity. It is common practice to blend different types of hops in a beer recipe.

Bitterness is quantitively measured for both hops and for beers in IBU (*International Bitterness Units*). A pale lager has a bitterness of around 12 IBU, while a Stout can hit 80 IBU or even more. Another critical measurement for hops are *alpha acids*. When the wort is cooked these become *isohumulones*, which are responsible for the bitterness. Hops are also rich in *flavenoids* (see page 76).

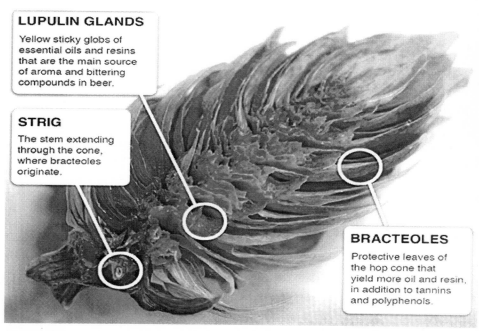

LUPULIN GLANDS
Yellow sticky globs of essential oils and resins that are the main source of aroma and bittering compounds in beer.

STRIG
The stem extending through the cone, where bracteoles originate.

BRACTEOLES
Protective leaves of the hop cone that yield more oil and resin, in addition to tannins and polyphenols.

Magnified View (actual size varies from 2 to 5cm (about 1-2 inches)

Beer Glasses

Most European breweries have their own glassware that they feel is optimum for their product. Taverns often have dozens of different glasses on hand, each for a specific beer. There are only three that are essential: A stein for lagers, a so-called "vase" for Hefeweizens and an Abbey goblet for high-octane Belgian ales.

Beer Dehydrates

One cautionary note. Although beer tastes very refreshing and is mostly water, the fact is that it takes even *more* water for your liver to process the alcohol in it. Plus there is a diuretic effect. In studies, the consumption of 200ml (7 ounces) of a lager with 4.5% alcohol (low) resulted in 320ml (11.3 ounces) of urine, for a net loss of water in the body of 120ml (4.3 ounces). The solution to this is to drink water with your alcohol. Dehydration is a major source of hangovers, too.

TIS THE SEASONING

Even though most cooks will recognize the names of the spices listed here, most people don't have a good understanding of what they really are, or when to use them.

ENGLISH NAME	FAMILY	NOTABLE FEATURE
Allspice	Myrtle	Actually a dried fruit. Used worldwide, especially Jamaica.
Bay Leaves (Laurel)	Laurel	Used in soups and braises. The flavor changes on long cooking.
Beau Monde	various	Propietary blend of celery salt, MSG, sugar, pepper, dried onion.
Bouquet Garni	various	Often sold in bottles, this is just a blend of assorted dried herbs.
Chervil	Apiaceae (Parsley family)	Essential herb in classic French cuisine, but otherwise rare.
Chili Powder	Genus Capiscum (Nightshade family)	Usually a mix of dried chilies with cumin, oregano and more.
Chinese Five Spice	various	Sichuan peppercorns, cloves, anise, fennel seeds, cinnamon.
Chives	Amaryllis (as are onions and garlic)	Distinctive onion-like flavor vital to French and Swedish cuisine.
Cinnamon	Laurel	In western cooking this means Cassia and not true cinnamon.
Cloves	Myrtle	Widely used around the world, but rarely as a primary flavor.
Juniper Berries	Cypress	Especially used in Scandinavian cuisine and with game meats.
Lovage	Apiaceae (Parsley family)	Used in Europe for centuries but now mostly forgotten.
Mustard	Brassica (Cabbage)	Dry mustard is yellow but there are several types of seeds.
Old Bay Seasoning	various	Propietary blend of ground bay leaf, celery salt, pepper, paprika.
Poultry Seasoning	various	Blend of herbs, especially thyme sage, marjoram and rosemary.
Pumpkin Pie Spice	various	Blend of dried ginger, cloves, cinnamon, nutmeg and allspice.
Sorrel	Polygonaceae (Buckwheat family)	Mostly used in soups, especially in eastern Europe.
Star Anise	Schisandraceae (related to Magnolia)	Powerful spice utilized in molecular gastronomy.

* Additional Information *

Allspice: Despite the name, this is not a blend of spices. This is the dried berries of a tree first brought to Europe by Christopher Columbus from his voyage to Jamaica. In America this is rarely used by home cooks in anything but desserts, but it is a vital savory spice in other cuisines (especially the Caribbean and the Middle East). It is also a key ingredient in many sausages and commercial barbecue spice mixes. It has more potential to elevate savory foods than most western cooks consider.

Bay Leaves: There are two products sold as bay. Namely Turkish bay leaves and California laurel, although the latter is mostly common in the United States (especially when you see fresh bay for sale). The two are not botanically related. They can be distinguished by the shape and color. Turkish bay leaves are generally lighter green and more round in shape, as shown in the photo below. Not all leaves obey this rule. Judge by the average.

Turkish Bay Leaf

California Laurel (Bay)

Just about every soup and stew recipe calls for bay, and yet most cooks have no idea why. The reason that this is perplexing is because raw bay is very different from what it smells and tastes like after long slow cooking, and the final effect varies depending on what it is being cooked with. Also be aware that the eucalyptus aroma and taste in California bay will persist after cooking, which is generally considered a bad thing in European cuisine. However, this can be desirable in other cuisines, especially in some Mexican dishes and in the curing of meats. The chemistry of how Bay works is involved is explained in more detail on page 53.

Beau Monde: This famous blend is often called *Beautiful World*. It's been used for generations but few realize that the commercial product is mostly just MSG, salt, sugar and onion powder made from toasted onions. It compliments anything savory, as you would expect. If you like this seasoning, then try making your own more sophisticated version using the *Beautiful World* recipe on page 233.

Bouquet Garni: This is a frequently seen blend of herbs, sometimes ground to a powder. It is marketed as a cheap and easy replacement for a true *Bouquet Garni*, which is a little bundle you tie up with string (or put into a cloth bag) that generally includes bay leaves and thyme, and may contain many other fresh herbs as well as peppercorns, juniper berries and more. The commercially bottled product is a very poor imitation of the real thing that's most often used by the likes of delicatessens and caterers.

Chervil: One of the four most important herbs in classic French cuisine (the others being thyme, tarragon and chives). Yet chervil is almost never seen in the cuisine of any other country. Just as lovage can be compared to a cross between parsley and celery leaf, chervil can be compared to a cross between parsley and anise. However neither of these descriptions are completely accurate. Each has its own unique flavor. Outside of France, the dried product is the most common type found. Unfortunately, like other delicate herbs such as tarragon and dill, the dried product is distinctly different from the fresh. That's not to say that it isn't useful sometimes, but the high aromatic notes that make chervil prized in classic French cuisine are distorted in the dried product.

Chili Powder: Although occasionally you find bottles of a single type of chili pepper that has been dried and ground, most of the time "Chili Powder" should really be labeled, "Spice Blend for Making Chilli", as in *Chilli Con Carne*. Sometimes chili peppers are not even the principle ingredient! If you want to add dried chilies to something, then dry and grind some yourself. Otherwise make your own Chili Powder (see Volume 2, page 192). There is also a more complex specialty version, *Texas BBQ Chili Powder,* on page 235 of this volume.

Chinese Five Spice: This blend is traditionally comprised of roughly equal parts of star anise, cloves, cinnamon, fennel seeds and Sichuan peppercorns. This is one of the spices that can be found on the shelf of most serious cooks, but shouldn't be on the spice rack of *any* serious cook. The reason is that this ground mixture degrades very quickly because of reactions that take place between the volatile flavor compounds at room temperature. It should always be mixed fresh and used within a few days.

Chives: There are both onion and garlic chives, but the former type is what we normally refer to as "chives". They have a distinctive flavor and are not a substitute for either onion or garlic in any dish. The freeze-dried product is the best bottled dried herb there is for resembling the taste of the fresh herb. For more, see pages 73-74.

Cinnamon: With the exception of Indian cuisine, "cinnamon" is taken to mean the bark of the Cassia tree. Like Allspice, cinnamon is primarily used as a spice in desserts in America and Europe these days. Yet it was very popular for savory dishes in the past and continues to be commonly used with meat and vegetables in Mexico, China and the Middle East. It is a very good spice to add in small amounts, but you must be careful or the guest's mind will fill with visions of apple pie.

Cloves: Here is another spice that has waned in popularity as a savory ingredient in the west. Although cloves remain an important part of French cuisine (especially as *piques*, which are clove-studded onions for stocks and certain sauces). Cloves are very often overlooked as a background note by less experienced chefs.

Juniper Berries: This is another example of a spice that is not well known to less experienced cooks. The flavor can be described as a cross between rosemary and cranberries with a hint of pine nuts. Although mostly popular in Scandinavia, juniper is commonly used in making sausages and cured meats throughout Europe. It is also used in many cabbage and sauerkraut recipes. Juniper is a natural to pair with any wild game meat or wild fowl except duck, which it clashes with in an *anti-resonant* way that is not beneficial (see Volume 2, pages 25-26). Juniper Berries need to be crushed to release their flavor, but because they will not dissolve or break down much during cooking you must either strain them out (even if you have blended a sauce containing them) or you will be left with unpleasant grainy nuggets in the final dish. Sometimes they are left whole in pan fried dishes for the diner to either eat or pick out.

Lovage: This is one of the oldest herbs used in cooking, and at one point was as common as thyme and parsley throughout Europe. It remains very popular in Russia and parts of Eastern Europe, but it is not sold commercially (either fresh or dried). It is an herb that people grow themselves in their dacha, or summer house. You can purchase it online from gourmet spice merchants, and I strongly urge you to do so. The flavor is unique and highly useful. Although you can substitute equal parts of dried celery and dried parsley in recipes, it's a pale comparison to the real thing.

Mustard: The condiment you get in a bottle is generally a mix of ground mustard seeds, vinegar, salt and other ingredients. Most American mustards contain a lot of turmeric, and if that's the flavor you grew up on, you may find European mustards to be strong and even a bit strange tasting. Dry powdered mustard is generally preferred in cooking because you know exactly what you are getting. Coleman's is an excellent brand and commonly available.

Old Bay Seasoning: This is a proprietary blend. The main ingredients are ground bay leaves, celery salt, pepper and paprika. There is also dried ginger, mustard, nutmeg and a few other minor components. It is most frequently used in poaching broths for seafood, especially shrimp. It has a distinctive taste that many Americans grew up with, so they associate the taste of seafood with this spice and for many, shrimp don't taste like "shrimp" without it. The fact is that it easily overpowers natural flavors and excess bay leaf is a fast path to unexpected bizarre flavors (see page 53).

Poultry Seasoning: While there are some variations between brands, the general formula is dried thyme, sage, rosemary, sage, salt and pepper. MSG is almost always present, though seldom stated as such, instead being listed as *autolyzed yeast extract* or *natural flavors*, to pacify consumers who are fearful. The problem is that it loses flavor quickly, and your bottle is probably years old.

Pumpkin Pie Spice: This spice wins the prize for being most likely to be in your spice rack and tasting like dust because most people only make pumpkin pie once or twice a year (if at all) and they have no use for it any other time. The mixture of dried ginger, ground cloves, cinnamon, nutmeg and allspice is simple enough to add as individual components with far better results, so unless you own a bakery that makes pumpkin pie, there's no reason to buy this.

Quatre épices: Uncommon in modern recipes, but one of the main seasonings used in French cuisine in the past. As you might have guessed, it means "four spices", and it is a mixture of ground peppercorns, nutmeg, cloves, and ginger in the respective ratio of 5:2:1:1. These days black peppercorns are usually used in commercial bottlings, however, the original formula for this was based on Grains of Paradise (see page 58). Considering that the pepper is over half of the mixture, obviously such a product is not authentic. Some manufacturers add allspice to a mixture of black and white peppercorns to try and compensate, but it is not the same. There is even a brand that includes cinnamon, but no pepper at all because it is intended for use in desserts.

Sorrel: Also known as Spinach Dock. Used primarily in soups, the taste has been compared to a kind of herbal kiwi fruit with a hint of strawberry. It's mostly popular in Russia, Ukraine, Romania, and Poland, although it also shows up in a few dishes from Hungary, Croatia, Greece and Albania, where it's also primarily used in soups. This plant is rich in oxalic acid, which is toxic, but small amounts are safe unless you have kidney problems. Before you worry too much, consider that apricots and cherries contain small amounts of cyanide and we eat those, too. Sorrel is also used in some of the spice mixes I've presented starting on page 217.

Star Anise: Botanically unrelated to anise, but it does contain the same key flavor chemical, *anethole*. This spice has gained notoriety in the west in recent years after it became a favorite among some pioneers in molecular gastronomy. In very small amounts it can give beef and pork more flavor on a subconscious level, but the key here is *very* small amounts. It is second only to nigella for having the ability to shine through every other seasoning. Like a lighthouse on a foggy night at sea, this will be the only thing you see if you use too much. This is not as much of a problem in dishes of India and Southeast Asia (where it is especially popular) because of the massive amounts of other seasonings and hot chilies that can stand up to it. In European cuisine, a little goes a very long way.

PORCINI & TRUFFLES

These culinary jewels are problematic for various reasons.

Dried vs. Frozen Porcini

Although fresh is always the best, it is a grim reality that wild mushrooms such as Porcini are only in season a short time each year, and only in a few small regions of the entire world. The rest of the time your only options are frozen or dried.

Whether you reconstitute dried mushrooms by allowing them to stand in warm water, or you defrost frozen mushrooms, you end up with a mushy sponge that can't be pan fried in the way you would have been able to do if they had been fresh. This is important because the greatest beauty of these mushrooms is only achieved when they have been browned.

There is one partial solution to this dilemma if you have access to fresh wild mushrooms at any point in the year. Namely, you fry them in oil and/or butter when they are still fresh and then freeze them for later use. When you defrost them, the flavor will still contain at least some of those caramelized notes.

The stringy, goopy texture is a problem, though. My solution to this is to cook your porcini (reconstituted from dried or frozen) into a base sauce of some sort for the intended dish. Then fry ordinary champignon mushrooms for the texture and combine them with this base sauce for the additional flavor it brings. This is what I did in the *Veal Scallopini* recipe (page 174). The downside is that it still doesn't look very good aesthetically, but sometimes you can conceal this, such as in a pie or the filling for ravioli.

Porcini Scams

As with other expensive products, there are unscrupulous sellers who offer cheap imitations. There are many types of porcini-like mushrooms that are edible, but the real ones have an extremely strong aroma. If there is very little aroma, they're fake. There are some companies selling porcini pastes of various types in jars. Most of them are only a tiny fraction of actual porcini and the rest are fillers of different sorts.

Black Truffles

When it comes to black truffles, it is fortunate that they can be bottled so as to retain a lot of the aroma and flavor. Beware of imitations from China, though. Not long ago a fungus was discovered in China that looks exactly like a black truffle, but it has virtually no aroma or taste. These are often bottled and marketed as black truffles at low prices. You get what you pay for. If it is cheap, it is almost certainly a fake. The Chinese ones have even less flavor than an ordinary mushroom. A reputable brand of bottled black truffles is Giuliano Tartufi of Italy. Once you open the bottle you need to use them in two weeks.

White Truffles

White truffles are more perishable. They can be packed in rice for a short time, but they are prone to rotting. The best solution to introducing white truffle flavor into a dish at a reasonable price is by using a cream of white truffles that comes in toothpaste-like tubes. The flavor is excellent and completely natural.

Crema de Tartufata Blanca

(Cream of White Truffle)

The tubes are expensive, but since you only need a smidgen in most cases, they are very cost effective. It is also sold in glass jars, but I recommend against this because that product spoils. The tube seals out air so it has a long shelf life if you keep it refrigerated. Avoid products that contain cheese or other filler ingredients.

Finally, truffle oil is not always just a synthetic chemical flavoring. Don't believe the stories you might have heard. There are real truffle oils made from fresh truffles—but they are expensive and perishable.

COMMERCIAL PAPRIKA

In previous volumes I extolled the virtues of oven drying red peppers to create your own paprika. However, there are varieties of commercial products that you should be familiar with, and sometimes must be used to obtain authentic flavor.

HUNGARIAN

Különleges: Very mild flavor. Seldom useful in my opinion.

Csípősmentes Csemege: Not spicy but rich in flavor. This is the type you produce using the method in Volume 1, pages 23-24.

Csemege Paprika: Stronger in aroma than in taste. Best used in dishes where the paprika is added at the end.

Rózsa: Rather the opposite of Csemege, being stronger in flavor but milder in aroma. The bright red color is an important aspect for the aesthetics of many Hungarian dishes.

Édesnemes: This is the generic exported "Hungarian Paprika", and has no dominant character. It's red and flavorful, but specific brands can vary, especially with age.

Félédes: The same as Édesnemes for all intents and purposes, even though it is produced in a different way.

Erős (Hot): This is a useful ingredient that has no exact equivalent from any other country or production method. The best you can do to substitute for it is a mix of cayenne and regular paprika, but get the real thing if you can. Szeged is a famous brand of this.

SPANISH SMOKED PAPRIKA (PIMENTÓN)

Dulce: Like all Spanish paprika, this is smoked. It is the sweetest, and the default choice whenever Pimentón is called for.

Agridulce (Bittersweet): A sharper and more metallic flavor. Not as useful as either Dulce or Picante in my opinion.

Picante (Hot): Aside from being smoked, this is quite distinct from the Hungarian hot type. It is brown in color and the strongest of all.

THE LESS DEADLY NIGHTSHADES

It is easy to forget that these diverse plants are all members of the same botanical family called nightshades. If this is news to you, then it's probably quite surprising. What's even more surprising is that there are actually very few members of this family that are edible. Most of the nightshade family is deadly poisonous.

Potatoes

Eggplants

Paprika

Tomatoes

Peppers
[Sweet and Hot]

Even within this group there are some dangers. Potato leaves and green spots on potatoes <u>are</u> toxic. Do not consume them. On the other hand, the stems and leaves of tomatoes are commonly believed to be toxic, but they are perfectly safe—at least in reasonable amounts. Adding **tomato stems** to a slow simmering tomato sauce is a great way to enhance its complexity. Treat the stems and leaves as an herb.

SCALLOPS & SHRIMP

Wet and Dry Scallops

Almost all of the scallops sold have been soaked in a solution of sodium tripolyphosphate. This is a preservative that also causes them to absorb a great deal of water. Since you are paying for them by weight, it makes for a higher profit margin since up to 40% is now water. Unfortunately, it also seriously damages the texture and flavor of the scallops. When you try to fry them, they are sitting there boiling in the phosphate chemical solution on the pan. You can help this out some by soaking them in several changes of water and then drying them on towels, but even so they are best suited to soups and dishes where they will be poached in a sauce.

"Dry scallops" (the industry term meaning that they have not been soaked in phosphates) are far more expensive. These are only in large cities and sold under the names of *day boat* or *diver* scallops. These terms mean that the boat that caught them returned to port the same day so that the fish did not need to be processed onboard. The vast majority of fishing is done from ships that stay out at sea for weeks, so they have to freeze and preserve their catch so that it won't spoil before they get back to port.

The Illusion of "Fresh" Shrimp

Most people mistakingly believe that the shrimp they see for sale in the display case of their local fish market are fresh. The fact is that almost all shrimp are frozen onboard the ship they were caught on and stored in blocks. The store defrosts these and puts them out on ice to save you the trouble of defrosting them yourself, but the reality is that they might have been defrosted several days ago and left sitting there until you came along and thought you were buying fresh shrimp. Nearly all restaurants buy frozen shrimp and defrost them in warm water on demand. This results in the best quality product at a reasonable price. Unlike scallops, there is not as much difference between fresh and frozen shrimp.

WHEAT FLOUR & GLUTEN

Wheat is one of mankind's oldest agricultural crops, and consequently—like beer and wine—has evolved into a topic that is quite technical with literally tens of thousands of different commercial species, numerous distinct farming methods, various milling methods and different treatments of the milled product, each resulting in flours with measurably different properties that even most professional bakers are unaware of.

The first thing you should know is that white flour is not necessarily bleached. The endosperm of wheat, which is where

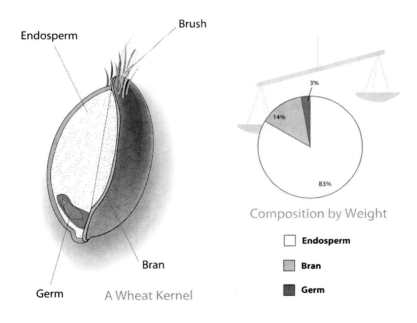

Endosperm

Brush

3%

14%

83%

Composition by Weight

☐ Endosperm

▨ Bran

■ Germ

Bran

Germ   A Wheat Kernel

most of the mass is at (see diagram above) is naturally white in most commercial types of wheat. So long as it is milled in such a way as to exclude the germ and bran there is no need to bleach it to make it white. However, this results in a lot of waste, so most manufacturers mill more than just the endosperm. The problem then is that the resulting flour has brown specks through it, so they bleach the flour to conceal the color. While the bleaching destroys some nutritional value, there's a bigger problem. Those little specks are still there. They are hard and have sharp edges. If you are making baked products such as bread, they cut holes in the gluten, resulting in less rising and other complications. *King Arthur* brand flour is often used by pro bakers because it is only the endosperm.

Gluten

The word *gluten* has became part of our everyday language, yet most people still don't really understand what it is because the concept is confusing. The endosperm in wheat contains 7-15% proteins (see page 64). These protein molecules unfold as long strings when they are wet. When you knead dough, you are tangling up these strings into nets. The nets are able to trap carbon dioxide gas bubbles released by yeast (or from chemicals such as baking powder) and provide a rise in breads, cakes and other baked goods. The more dense the network, the firmer the product will be and the more rise it will have. Pizza crust is on one end of the spectrum and angel food cake is on the other, but they both rely on gluten to trap carbon dioxide. The highest protein of all flours is Hard Red Spring Wheat, which is about 15% (see chart at right).

That's what gluten is—tangled webs of proteins, but not just *any* protein—some grains have protein that does not form gluten (see the table on page 88). The proteins in the endosperm of the wheat berry are about equal parts *gliadin* and *glutenin*. Technically, these

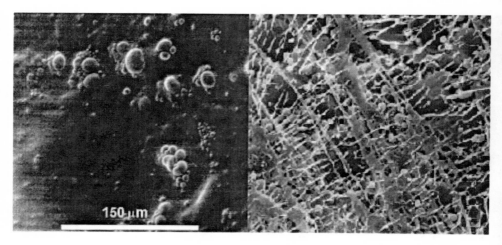

Electron micrographs of freshly wetted flour (left) and kneaded dough (right).

Enriched	Nutrients have been added to the flour. This is usually to compensate for the loss during processing / bleaching. Typically iron and B vitamins are added and sometimes calcium. It is still less nutritious than unbleached flour.
Bleached	Sometimes this applies to red wheats, but usually it is the chemical treatment of speckled wheats that were milled with the bran for higher productivity.
Self-Rising	Baking powder has been added to the flour. The problem is that you have no control over how much or what kind.
Durum	Very high protein (13-14%) species of yellow wheat that is mostly used in making pasta. See Semolina below.
Semolina **Farina**	Applies to any coarse milled wheat. Usually durham wheat that is used to make pasta, but not always. When it is a white wheat, it is properly called **Farina**.
Wondra Flour	Low protein flour cooked with steam and dried. Malted barley is added. Used in sauces, gravies and baking.
Red Wheat	More assertive in flavor than white wheat. Although it is commonly a hard winter wheat, there is also soft type.
White Wheat	Sweeter and milder in flavor than red wheat. This can also be hard or soft, winter or spring. It is naturally white without bleaching, but sometimes it is bleached anyway.
Winter Wheat **Spring Wheat**	This refers to the winter planting of some hardy wheat species in some US states. The wheat survives under snow to become more flavorful. Similarly, **Spring Wheat** is grown in other states, and both may be red or white.
Hard Wheat	This only means it has a high protein content (12-15%). This can be red or whiite, and grown any time of year.
Soft Wheat	This only means it has a low protein content (7-9%). This may be red or white, and grown any time of year.
Cake Flour	Finely ground bleached soft winter wheat. Sometimes imitated by adding 10% corn starch to regular AP flour.
All-Purpose Flour **"AP Flour"**	Mid-range protein content (10-11%) and usually sold as both bleached and enriched (see above).
Whole Wheat	Generally made by milling whole red wheat berries, yielding about 14% protein, but brands vary (see text).
Bread Flour **(Bread Machine)**	This is a high protein flour that's supposedly optimized for making bread, but like "whole wheat" it varies widely.

are not specific proteins, but rather classes of proteins—but you don't need to concern yourself with that to understand the process. Gliadin is soluble in water, and it is the protein that causes reactions in those with celiac disease because it can cross the intestinal barrier and enter the bloodstream. Glutenin is not soluble in water. Once water is added to flour, these proteins begin crosslinking into gluten. The crosslinking is not rapid because of the insolubility of glutenin, however if the dough is left to stand for a long time, gluten forms by itself. This is the basis of "no-knead" breads. The addition of fat to dough interferes with gluten formation and leads to flakier and lighter products like cakes and pie crusts.

Celiac Disease

Before you jump to the conclusion that you have celiac disease, consider two things: Even the largest organization that exists to collect money for researching a cure for this disease admits that less than 1% of people actually have this condition—and that it is notoriously one of the most difficult diseases to accurately diagnose even after you have an intestinal biopsy! An entire industry has sprang up to market gluten-free products as a diet fad, and they aren't going to go away. This marketing has led to a general fear of gluten and people asking their doctor to be tested. But, because it can't be detected with certainty, the medical profession tends to err on the side of false positives in order to protect against potential lawsuits. I'm not saying that celiac disease doesn't exist, but I'm saying that for every person who actually has this rare condition, there are many more who are torturing themselves on a restricted diet purely out of psychosomatic hypochondria.

Gluten-Free Foods

To satisfy the demand for gluten-free products, many manufacturers have invented products with unusual chemical additives and novel materials that have not been extensively tested for their long term effects on humans. I remind you that for decades margarine was widely promoted as a healthy substitute for butter. Today we know that margarine is terrible for your health and butter really isn't bad (in moderation, of course). The same may well prove true of gluten substitutes in the future. In fact there have already been reports linking xanthan gum to birth defects according to the FDA. There is no way to be perfectly safe no matter what you do.

Potassium Bromate

Although it has been banned in many countries around the world as a proven carcinogen in rodents, the United States continues to allow potassium bromate to be added to flour. This is often called "bromated flour", and it has been in steady use since 1914. For this reason many American baked goods (including some potato chips and products that you wouldn't even expect to have flour in them) are banned in other countries—including China. Potassium bromate *can* be used relatively safely in the hands of professional bakers (which is why the FDA continues to allow it), but if the item is not baked long enough, or hot enough, or too much bromate was added initially, there will be a toxic carcinogenic residue. So unless you are a professional baker, just steer clear of any flour that contains bromate. Check the ingredients on your bread, too. Bromate is sometimes disguised as additive E924. Don't fear all *E series* numbers, though. Some are perfectly safe. For instance, E260 is just vinegar, E300 is Vitamin C, and MSG— the maligned whipping boy of all food additives—is E621 (as well as many other aliases invented to hide it from fearful consumers).

Whole Wheat

No legal definition exists for products labeled in this way. Most consumers take it to mean *Whole Grain* Whole Wheat, but that often not the case. In commercial baked goods it is rarely the case, in fact. It is also important to know that true Whole Grain flour goes rancid rapidly unless it is stored in the refrigerator or freezer. This is one reason why flours labeled "whole wheat" are often not whole grain—to increase their shelf life, the germ of the wheat is largely removed because that's where the perishable oils are. It's also where a lot of the health benefits are, so read the product label.

Baking Bread

The information in this chapter is important for all aspects of cooking, but advice about professional bread baking is beyond the scope of this book. If you are interested in the topic, I encourage you to go to the following website for expert advice:

www.thefreshloaf.com

BATTERS AND DOUGHS

Below is a quick visual synopsis of the differences between various classic batters and doughs.

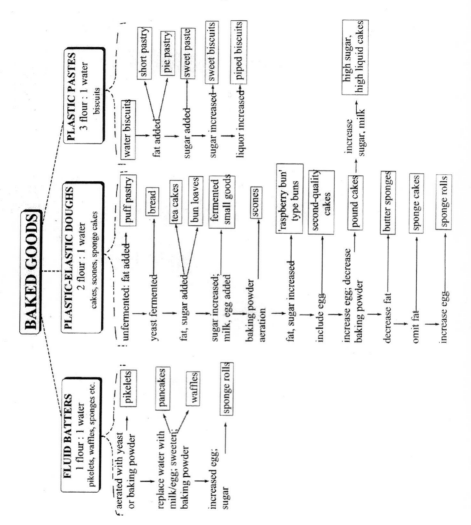

*Reproduced with kind permission from the class materials of the Swiss culinary academy, where I taught for a while.

ORGANOLEPTIC CHEMISTRY 101

In order to understand how flavors are altered and produced by cooking, you have to study the molecular nature of food. All plants and animals are comprised of cells. These cells have structural components (cell membranes) which are essentially flavorless by themselves because we can only taste and smell fairly small molecules. Even if you dried them and ground them into a fine powder, they would taste like gelatine. The reason is that you can't grind them anywhere close to the molecular size range, and our senses are only capable of tasting and smelling relatively small molecules. The bottom line is that when it comes to aroma and flavor, all of the players are fairly simple molecules, which is fortunate because that makes them much easier to understand. This field of study is called *Organoleptic Chemistry*.

Now you might think the Maillard reaction violates this principle. After all, when food is browned on the surface—the structural cell walls—we certainly get powerful flavors and aromas generated. Most of what you are smelling is the contents of the ruptured cells, though. A small amount of cell membranes do break off and participate in the reactions, but it's not the main source of flavor. See Volume 2 on bacon chemistry for some examples of specific molecules generated through cooking and how they are sensed.

All of the plants that we eat have a distinct aroma and taste without cooking them. This includes fruits, vegetables, herbs and mushrooms. There are two main reasons why plants produce these small molecules. Namely, either to attract insects for pollination or to repel insects and animals from eating them. Many of the flavors that we enjoy are deadly poisons to some insects and animals. One better known example of this is *theobromine* in chocolate. It is part of the flavor, being pleasantly bitter to our palate. It is also a deadly poison to many animals, including dogs and horses.

It is easy to get bogged down with the vast array of terminology involved in organic chemistry, especially when it comes to complex molecules such as terpenes and alkaloids, both of which are frequently key players in aroma and flavor. I am now going to attempt what might be impossible (we'll see)—to succinctly explain the key aspects of organic chemical reactions that affect flavors.

The first baby step on this rocket sled ride from, "I know nothing about chemistry" to having a working understanding, is to know how to read a structural diagram. Luckily, in food chemistry you only really need to know five elements and the "wildcard" symbol, **R**.

H = Hydrogen   **C = Carbon**   **O = Oxygen**

N = Nitrogen   **S = Sulfur**   **R = *Anything***

Keen in mind that **R−** is not usually a single atom, but rather *the rest of the molecule.* It might be anything from a hydrogen atom, or a single carbon atom with three hydrogens attached (called a *methyl group*) on up to a very long polymeric chain.

Because the vast majority of organic molecules are simply carbon and hydrogen, we use a shorthand form in which all unlabeled points are assumed to be carbon with any or all hydrogen atoms undrawn. How many hydrogen atoms depends on

how many other things are connected to that carbon atom. For the type of molecules we will be dealing with in food chemistry, every carbon has four bonds. Most oxygen and sulfur atoms have two

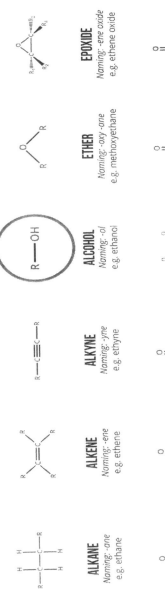

ALKANE
Naming: -ane
e.g. ethane

ALKENE
Naming: -ene
e.g. ethene

ALKYNE
Naming: -yne
e.g. ethyne

ALCOHOL
Naming: -ol
e.g. ethanol

ETHER
Naming: -oxy -ane
e.g. methoxyethane

EPOXIDE
Naming: -ene oxide
e.g. ethene oxide

HALOALKANE
Naming: halo-
e.g. chloroethane

ALDEHYDE
Naming: -al
e.g. ethanal

KETONE
Naming: -one
e.g. propanone

CARBOXYLIC ACID
Naming: -oic acid
e.g. ethanoic acid

ACID ANHYDRIDE
Naming: -oic anhydride
e.g. ethanoic anhydride

ESTER
Naming: -yl -oate
e.g. ethyl ethanoate

AMIDE
Naming: -amide
e.g. ethanamide

ACYL HALIDE
Naming: -oyl halide
e.g. ethanoyl chloride

AMINE
Naming: -amine
e.g. ethanamine

NITRILE
Naming: -nitrile
e.g. ethanenitrile

IMINE
Naming: -imine
e.g. ethanimine

ISOCYANATE
Naming: -yl isocyanate
e.g. ethyl isocyanate

AZO COMPOUND
Naming: azo-
e.g. azoethane

THIOL
Naming: -thiol
e.g. methanethiol

ARENE
Naming: -yl benzene
e.g. ethyl benzene

COMMON FUNCTIONAL GROUPS IN ORGANIC CHEMISTRY

bonds. Every nitrogen atom has three bonds. Hydrogen only has one. Just to be technically correct, I'll point out that the rules are actually more complicated, but not in the case of most simple food-related substances that are relevant here.

The physical nature (including the aroma and taste) is not just determined by the number and type of atoms, but also how they are arranged. Shown below are four types of butyl alcohol. *Butyl* refers to four carbon atoms. The **–OH** group makes it an ALCOHOL, as you can see from the chart on the previous page. Note that **R–OH** on this chart describes any alcohol, unless **R** is hydrogen, which makes it water. I'm not going to confuse you by explaining how each of these gets its name, but from the table below you can see that they are very different in their nature, despite all having four carbons and one alcohol (**–OH**) group. A tiny molecular change can result in a big change in the properties that we see, smell and taste.

	NAME	MELTING POINT BOILING POINT	
~~OH	1-Butanol	–90° C 118° C	Smells like banana and alcohol. Tastes faintly of vanilla in low concentration.
OH	2-Butanol	–115° C 100° C	Smells musty and ethereal. Taste is ethereal.
OH	iso-Butanol	–108° C 108° C	Smell is like high alcohol wine. Taste is whiskey in low concentration.
OH	tert-Butanol	25° C 83° C	Smells like camphor. "Indescribably horrible taste that lingers."
OH	Cyclobutanol	–91° C 124° C	Odor of roasted meat. Highly toxic - can't be used in foods.

Incidentally, butyl alcohols are rarely used as food flavorings because they are not very potent as such. However, they are often part of ESTERS (see chart on page 47). Esters are the product of ALCOHOLS and CARBOXYLIC ACIDS (again, see chart on page 47) with the loss of a water molecule. The chemical reaction for this is...

1-Butanol + Butyric Acid $\xrightarrow{- H_2O}$ Butyl Butyrate

Butyric Acid has the aroma of butter (which is where the entire butyl series gets its name from), but the product Butyl Butyrate has the aroma of pineapple. It is used commercially as a chemical additive in fruit snacks and drinks, but is also found naturally in fruits including berries, bananas, apples and more. If the same reaction is carried out with 1-Pentanol (five carbons instead of four) and Butyric Acid, the product is the ester, Pentyl Butyrate, that has the aroma of apricots with a high note of pear. This is another common artificial food flavoring, and also a flavoring in cigarettes!

This type of reaction—where a molecule of water (H_2O) is driven off to form an ester from an alcohol and a carboxylic acid—is called a *condensation reaction*. It is not always between two different molecules, though. If a single molecule has both of the required functional groups, then it can form a ring, such as the lactone shown below. Many lactone flavor molecules such as this one have coconut-like aromas and are used in cosmetics as well as foods.

γ-Nonalactone

What's important to keep in mind is that under the right circumstances many esters can undergo a reverse reaction called *hydrolysis* in which they break apart again back into their original acid and alcohol components. Now you can see how this would dramatically change the flavor. This is especially prone to happen when there is a large volume of water and an extended period of cooking, such as a stew. This is one reason why braises are generally more flavorful than stews or crock pot dishes (remember the rules for what constitutes a braise from Volume 2).

REACTIVITY

In addition to the hydrolysis type of degradation described above, there are many other reactions that can alter or eliminate flavors. Some molecules are more prone to reacting than others, though. The rules that determine which are likely to react and under

49

what circumstances is far beyond the scope of this book, but so as to not leave you hanging I'll explain a couple of the simpler rules.

Alkanes, alkenes and alkynes are odorless and tasteless by themselves. Alkanes are unreactive and alkynes are almost nonexistent in food chemistry, so we can ignore these two as players. Alkenes are reactive and can lead to rather surprising changes to molecules, especially when they are situated adjacent to other functional groups. The migration of electrons to an adjacent electronegative atom (oxygen, nitrogen or sulfur) is the basis for an entire encyclopedia of potential reactions—literally.

TERPENES

One of the most important categories of flavor molecules are the terpenes. The rules for what defines something as a terpene are complicated. Some examples are shown below, so it is easier to see the similarity. Note that Linalool does not have a closed ring, but it is still a terpene. These are the primary flavors of cardamom.

α-terpinyl acetate   1,8-cineole   Linalool   α-terpineol   Methyl eugenol

The rules for naming a terpene are also complicated and not important for you to know. What is important to know is that terpenes are generally powerful flavorings, chemically reactive and insoluble in water. They are soluble in alcohol, though. This is one of the reasons why wine and spirits are used in cooking—as a way of bringing terpenes out so that they can be tasted. Even though the alcohol is usually evaporated off before the food is consumed, the period of time when the alcohol was present will extract flavors efficiently.

Not only do terpenes react with other flavor molecules, but they are prone to changing just by heating them in the presence of oxygen. An example of that is shown here...

LINALOOL + O_2 → HYDROPEROXIDE FORM

The hydroperoxide form of linalool is a contact skin allergen. This has became a problem for the perfume industry where linalool is a common ingredient. Only a small amount changes to the hydroperoxide form. Just because a reaction *can* happen doesn't mean it will happen. Heat is the main factor, so keep your perfume cool.

CARDAMOM, BERGAMOT AND BAY LEAVES

On the previous page are the structural diagrams for the terpenes that are responsible for the primary aroma and flavor of cardamom. As in the examples cited in Volume 2 for bacon and tomatoes, none of the individual components smell like cardamom. The organoleptic properties are the result of a kind of musical chord of flavor molecules playing in harmony, as you can see in this chart:

COMPOUND	FUNCTIONAL GROUP	% IN CARDAMOM	ORGANOLEPTIC PROPERTIES
α-terpinyl acetate	Ester	34.6 – 52.5	herbal
1,8-cineole	Ether	23.0 – 51.0	camphor
Linalool	Alcohol	1.4 – 4.5	floral citrus
α-terpineol	Alcohol	1.4 – 3.3	sweet lilac
Methyl eugenol	Ether	1.2 – 1.4	tea; earthy

Note the large range in which the flavor molecules can exist. Cardamom is especially known for being highly variable, as mentioned in Volume 2, page 35. Commercially both the pods and the oil have five grades ranging from "very bold" to "weak", making it problematic in recipes where the amount specified can be too much if you happen to have a potent batch. You always have to adjust the amount of cardamom based on the nature of what you are using. The same is often true of other spices, too—although cardamom is especially extreme in this regard.

The next issues that comes into play are the lengths of cooking time and storage time, because reactions continue to take place in the refrigerator. If a dish is not strongly flavored when you put it in cold storage, it will come out bland a few days later. This is why my recipes are usually designed to produce concentrated flavors, since they were developed for restaurants where the dish would be reheated days later, usually with some water diluting it, too.

Just heating terpenes—and many other reactive flavor molecules—causes reactions to occur that alter the balance of flavor. You may not have heard of Bergamot before, but you have probably tasted it as one of the "natural flavorings" added to many commercial food products. I'm not talking about the Italian citrus fruit by the same name, but rather the plant that's a member of the mint family (*Monarda fistulosa*). It is not something you are likely to find in any supermarket, but it is an important ingredient in manufactured foods. Bergamot illustrates how terpenes change just on simple heating by themselves (let alone the myriad of reactions that occur in the presence of other substances). Below you see samples of the same Bergamot oil examined before and after it was

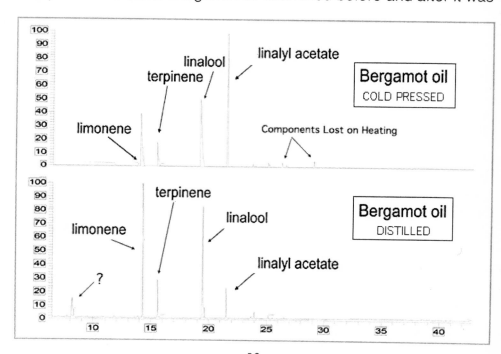

distilled. If it had not reacted then these charts would be identical.

The odor after distillation is still recognizable as Bergamot, but it s not the same. This is the same reason why you don't want to use he finest quality artisanal olive oils for frying foods, because the nice fruity flavors you are paying for will be distorted after heating. The best quality olive oils are reserved for cold dressings.

Bay Leaves are among the most mysterious of all the herbs because they change dramatically on extended cooking. It is important to note that there are two types of bay used in cooking. The Turkish variety, which is true bay, and laurel, a plant that is only vaguely related botanically and grown primarily in California. In almost every instance, the Turkish variety is what is called for and expected in recipes when it says "bay". One of the interesting things about bay leaves is that they break down and release *soprene*, which is the fundamental building block of terpenes. Isoprene is reactive and can combine with other available substances and oxygen from the air to produce a vast array of compounds that have flavors and aromas in tiny amounts, which add character and nuance to foods.

ISOPRENE

This process takes place slowly, and the results are largely random, so if you use too much bay, you can end up with something foul tasting. For example, *indole* is what is largely responsible for the odor of feces, yet it is an important ingredient in the perfume industry because a tiny amount is not perceived the same way that a large amount is. The same thing is true for the products that bay leaves (read isoprene) will produce. A little will add interesting character, but a lot can ruin a dish.

AUTO-OXIDATION

As explained previously with butanol as an example, aroma and taste are strongly influenced by the functional group(s). The same carbon chain will be smell and taste completely differently if the functional group is an aldehyde instead of an alcohol, for instance.

Now here's why this is so important: *Auto-oxidation.* Many flavor molecules react with oxygen in the air, either when heated or in the cold over long periods of time. For example, alcohols can be oxidized to aldehydes, and aldehydes can be oxidized to acids.

First Reaction (slow)

R—C(=O)(H) + O=O = R—C(=O)(O–H) + O–

| Aldehyde | Oxygen | Carboxylic Acid | Oxygen Free Radical |

Second Reaction (very fast)

R₂—C(=O)(H) + O– = R₂—C(=O)(O–H)

| Aldehyde | Oxygen Free Radical | Carboxylic Acid |

Autooxidation of Aldehydes to Carboxylic Acids

Understanding the exact mechanism for this requires an education in organic chemistry, but what you can see right away is that air and heat (or time) change the nature of flavors. Also, after the first reaction takes place, a highly reactive oxygen free radical is formed, which can act on a second aldehyde molecule or (because it is so reactive) many other substances that wouldn't react with oxygen alone.

Oxidation is usually a slow process at room temperature, but not for everything. Avocados, apples and artichokes all turn brown from oxidation rather quickly once they are cut. A purée of carrots doesn't take long to turn brown, too. Why brown? Because many different reactions take place producing compounds of different colors, and just like when you mix different color paints together, you get brown (more about this starting on page 59). But most of the time the effect is not visible as a color. Most changes that take place can be tasted and smelled, but not seen.

PEPPERCORNS

Next to salt, black pepper is of the most overlooked ingredients by home cooks. Conversely, the judicious use of pepper is a telltale sign of every experienced chef.

Black, green and white peppercorns all come from the exact same plant (*Piper nigrum*), which is native to South Asia. The spiciness of pepper is from the chemical *piperine*, and <u>not</u> from *capsaicin*, which is what makes chili peppers hot. The burning sensation on the tongue is vaguely similar, which is not surprising when you compare the similar structure of these two compounds, as shown below. Capsaicin is far more potent, though.

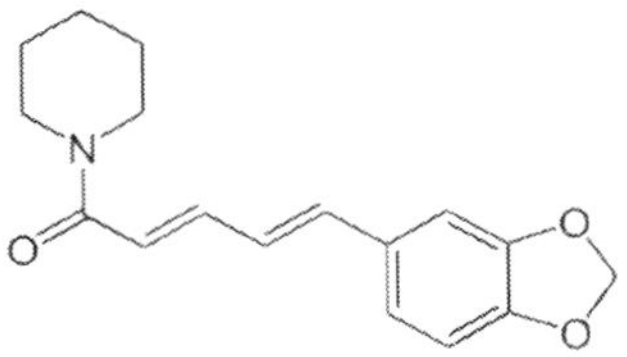

Piperine

Functional groups include an aromatic ring with two ethers (cyclic) and an amide. Total of 17 carbon atoms.

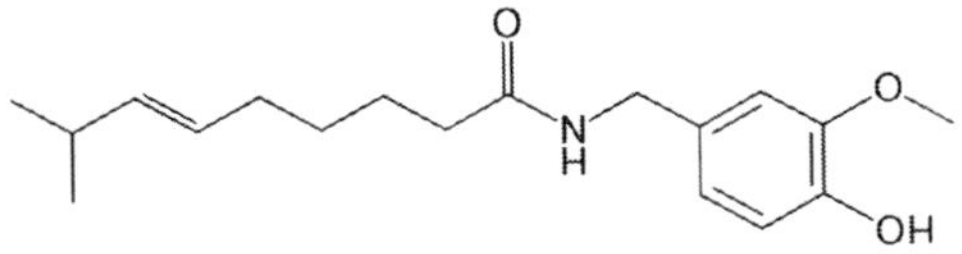

Capsaicin

Functional groups include an aromatic ring on an ether and an amide. Total of 18 carbon atoms.

You can even see some structural similarity to 6-gingerol, which is responsible for the spicy character of ginger and galangal.

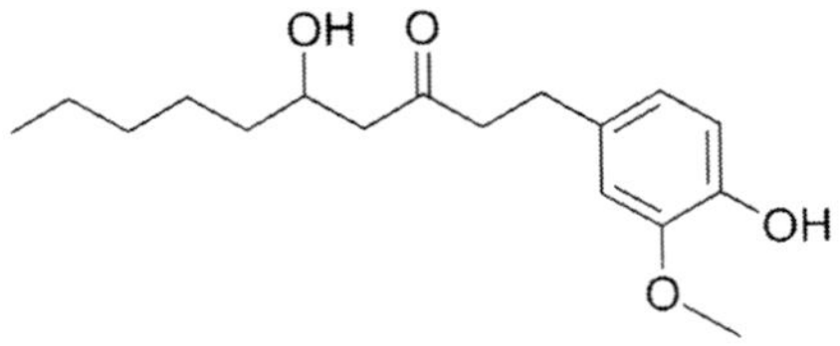

6-Gingerol

Functional groups include an aromatic ring on an ether, a ketone and an alcohol Total of 17 carbon atoms.

Keep in mind that all of these compounds react to some extent during cooking. The balance of flavors change as components evaporate and react. Depending on the amount of cooking and the temperature, the pungency is reduced or even lost altogether. The same is true for foods after prolonged storage, even if kept cold.

Black Peppercorns

The berries are picked while still unripe and green. Then they are sun dried where they shrivel and turn black.

Tellicherry black pepper comes from grafted plants grown on Mount Tellicherry in India, and is regarded as the best in the world. These days about a third of all black pepper comes from Vietnam, though. Some comes from Bangladesh.

Footnote: Those huge restaurant peppermills were invented so that a waiter could grind pepper from across large corner tables.

White Peppercorns

The berries of the plant are allowed to fully ripen until they turn red. Then they are put in cloth bags and floated on water for 7-10 days. The outer layer is stripped off to reveal the inner white pepper that is dried. Since much of the piperine is in the outer layer, white pepper has only about half that of black pepper.

Muntok pepper from Indonesia is generally regarded the best white pepper. Vietnam also produces white pepper.

White pepper is often chosen for aesthetic reasons in light colored dishes where flecks of black pepper would stand out. The flavor is quite different, though.

The table below summarizes some of the chemical differences between black and white peppercorns, as well as providing some

FLAVOR CHEMICAL	INDIAN BLACK PEPPER	BANGLADESH BLACK PEPPER	WHITE PEPPER
Piperine	7.5	4.5	6.0
β-Caryophyllene	18.4	10.6	16.0
α-Pinene	16.7	7.4	2.5
β-Pinene	13.6	13.2	7.3
Limonene	16.1	15.2	11.9
Linalool	0.3	0	1.5
Humulene	0	0.5	0.9
Myrcene	2.9	3.1	0.9
Bisabolene	0	0	7.4
Cubebol	0	0	4.4

insight into the considerable range between peppercorns from different sources.

The piperine content varies depending on the source, the age, and the method used to measure it. In general, black peppercorns have 5-10% piperine and white has less because piperine is largely in the outer membrane that's stripped off (as previously mentioned).

Green Peppercorns

These are the unripe drupes of *Piper nigrum*, the same plant that produces black and white peppercorns. Instead of boiling them first, they are dried when still green. They are also sold packed in brine, or pickled with vinegar without having been dried. The canned or bottled green peppercorns in liquid are typically used whole in sauces, especially for beef and game meats.

Like black and white peppercorns, there are variations in the chemical composition. Using dried green peppercorns in place of black is a good way to introduce a new dimension to a classic dish.

Cubeb Pepper

Another member of the pepper family, *Piper cubeba* mostly comes from Sumatra. The same plant grows in Java and is sometimes sold as Java Pepper, or "tailed pepper".

It has the pungent flavor of pepper, but with stronger notes of allspice and camphor than *Piper nigrum*. Cubeb Pepper is also used in flavoring cigarettes and gin.

Long Pepper

Named for the slender elongated shape, it resembles catkins. *Piper longum* is native to India, but a close relative, *Piper retrofractum,* from Java is sold by the same name. They are more pungent than black pepper, but with some sweetness as well as chili pepper-like notes. Very popular in Malaysia and Indonesia. It is also a common ingredient in the spice blends of North Africa.

Key References

"Comparative Studies on Physicochemical Properties and GC-MS Analysis of Two Varieties of Black Pepper (Piper nigrum)", Aziz, et.al., *International Journal of Pharmaceutical and Phytopharmacological Research*, **2012**, *2* (2), pp 67-70

"Chemistry, Antioxidant and Antimicrobial Potentials of White Pepper (Piper nigrum)", Singh, et.al., *Proceedings of the National Academy of Sciences, India Section B: Biological Sciences*, **2013** 83 (3), pp 357-366

NON-PEPPER PEPPERCORNS

Several dried berries that are often mistaken for types of pepper.

Grains of Paradise

The chemical component responsible for the peppery taste of this is *paradol*, which is extremely similar to 6-gingerol (shown on page 55), the chemical responsible for the mild burning sensation of ginger and galangal.

Paradol

Functional groups include an aromatic ring on an ether and a ketone. Total of 17 carbon atoms.

However, grains of paradise also contain a number of other substances that give them notes of citrus and spice that are unique.

Pink Peppercorns

These are from a Peruvian tree that is a member of the cashew family. The flavor is distinctive and the chemistry is not well understood. There are two principle components. One is vaguely similar to capsaicin and so it is believed the mode of heat is probably similar. The other component is *myercene*, which is a kind of "blank slate" terpene like the isoprene is in bay leaves (page 53). This enables pink pepper to react and form other complex flavor molecules, as in *Mafalde with Roasted Cauliflower* on page 182.

Sichuan Peppercorns

Dried berries of the prickly ash tree, which is native to the Sichuan province of China. The outer husk has all the flavor and aroma. It causes a slight buzzing sensation on the tongue, but the chemistry for this is not fully understood yet. Sichuan pepper is aromatic with a scent of anise, unripe flower buds and faint citrus. It is one of the components of Chinese Five Spice. Many Chinese chefs mix it with coarse salt and roast it in a dry pan before use.

Aleppo Pepper —See the notes on page 281.

COLOR IN FOOD

Color arises when portions of the visible spectrum are absorbed by molecules. Because the portion of the electromagnetic spectrum we can see is quite limited (see color diagram on the back cover of this book), most molecules do not absorb *any* of these particular wavelengths, so we see them as white.

The first three groups shown on page 47 are alkanes, simple alkenes and alkynes. These are all white (or clear if they're a liquid). All amino acids and simple sugars are white, too. Sugars turn brown when you cook them, but that's due to chemical reactions that take place to form more complex substances.

CONJUGATED ALKENES

In order for a molecule to absorb light within the narrow region of the visible spectrum that we can see, there must be a *low-energy electronic transition* possible. This usually involves several alternating (*conjugated*) alkenes. An example of this is *lycopene*, which is the red pigment in tomatoes, pink grapefruit and more.

Lycopene

Because there are eleven conjugated alkenes (carbon-carbon double bonds), lycopene absorbs most of the light from orange on through ultraviolet. But red is just a bit too low in energy for it to promote an electronic transition, so red is reflected, and that's why we see lycopene as red. If it didn't have quite so many conjugated carbon-carbon double bonds, it would be orange or yellow in color.

Higher energy wavelengths are more likely to be absorbed (less likely to be reflected). This is why blue and violet colors are rare in nature. In order for something to be blue or violet, it has to absorb *lower* energy wavelengths and reflect higher. Many colored food compounds are *anthrocyanins* (page 77). There are also...

PORPHYRINS

Red in blood is due to heme in hemoglobin. The green in leaves is due to chlorophyll. Both of these intense colors are the result of a porphyrin that holds (*sequesters*) a metal ion in the center. Iron for blood and magnesium for chlorophyll. There are also dozens of enzymes that have various metal ions sequestered in a porphyrin.

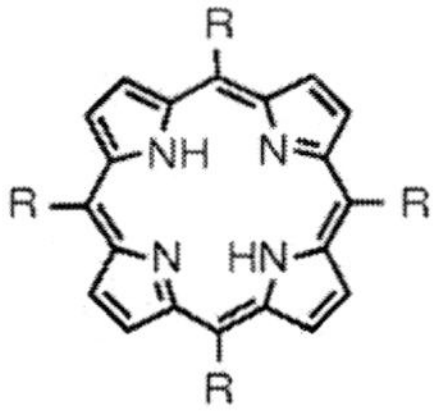

Porphyrin	Chlorophyll	Heme
(purple)	*(green)*	*(red)*

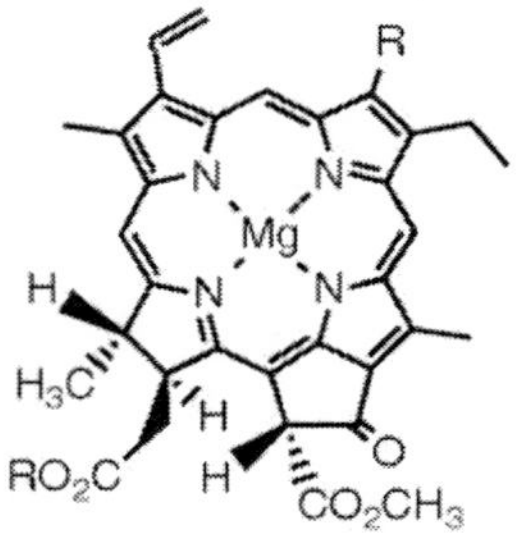

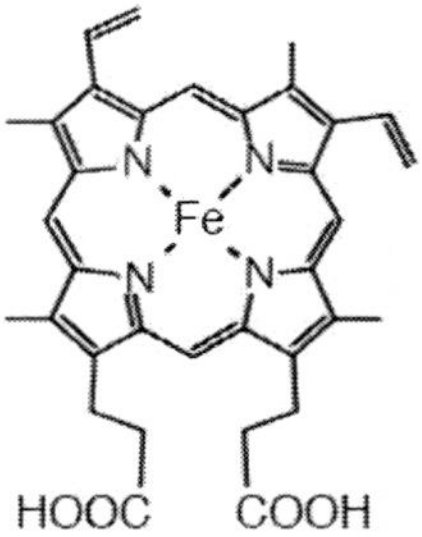

THE CURIOUS CASE OF TANNIC ACID

Tannic acid is a prominent flavor molecule in coffee and tea. Small amounts are also in red wine. However, "tannic acid" is not actually an acid, as you can see in the diagram below.

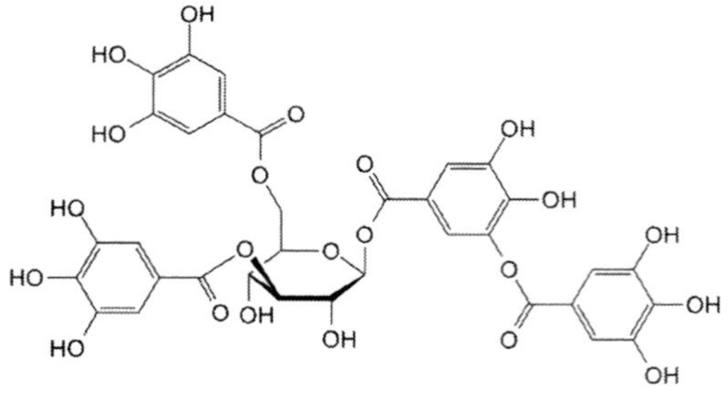

Structure of a Tannin

More important, it is not a single structure. It is a mixture of different related compounds called *tannins* with varied amounts of branching. This is why tannic acid is brown and not a prismatic color, just as you get brown if you mix different color paints together. Commercially it is often used as a brown stain. Tannic acid was isolated long ago when chemistry was still in its infancy, and the mistakes of that era have been passed down in the language. This mixture of colors is why brown food is brown, too.

▽

OLIVE OIL QUALITY

These days many people are leery of the olive oil industry. In a recent highly publicized study from the University of California Davis, dozens of olive oils sold as "extra virgin" in California supermarkets were tested both chemically and by two panels of tasting experts. The results were shocking—or at least that's how the abridged version of their report was presented to the public. Over two-thirds of the products tested failed to meet the (somewhat arbitrary) criteria for "extra-virgin". The conclusion drawn in the interpretation of these findings was that Italian olive oils are not to be trusted for purity. By means of press releases, online websites and paid advertising, the message was sent: The best source for extra-virgin olive oil is that made in California. Did I mention that a coalition of California olive growers funded this study? (They did.)

Your first clue that this was not entirely legitimate is the fact that it was not published in a scientific journal. Why wasn't it? Because such journals involve a rigorous peer review process by qualified scientists that this report could never pass. The bias in the study was compounded by the mass media retelling of the report with even less integrity. So now many people think that extra-virgin Italian olive oil is a scam—but this is not the case.

There are a many problems with this study. First, there were several testing methods involved. There was no clear distinction between imported and domestic olive oils on average according to most of the methods used. In fact, 100% of the bottles from both Italy and California passed three of the tests perfectly! Another test was only failed by some of the bottles from Pompeian, which was the least expensive brand tested.

Only one test showed a consistent difference, so they made a point of calling attention to that one, ignoring the fact that they passed other tests with flying colors. Their assumption that failing this one test was somehow proof of bad quality was never shown.

Still, the most dishonest aspect of the study was the consistent use of ambiguous language in which olive oils had failed *either* due to adulteration with other types of oil or had been subjected to heat during storage. At no point in their research was any oil shown to definitely contain soy oil or fish oil. Those were simply suggested as a *possible* explanation, but of course the media did not report it that way. In a court of law, this would be like the prosecution saying, "You didn't see your wife that day, which might be because you murdered her the day before." Sure, it *might* mean that, but there are a lot more likely and less sinister explanations. By offering that as a plausible explanation, the jury's mind has been poisoned in favor of an unsubstantiated claim. Lawyers don't usually get away with such tactics in court, just as studies of this sort don't get published in scientific journals.

If you still aren't convinced, then consider this damning evidence: If the oil had actually been adulterated by the manufacturer, then we would expect to see 100% of the bottles from that producer fail all of the testing, but what we see are typically only some bottles failing to meet their criteria. That alone proves that there is no intentional adulteration, because it wouldn't happen selectively on some bottles but not others.

The last thing to consider in this report is the taste testing. Very little information about this process or the results appears in the version of the report that is circulating online. What we do know is that some of the oils that performed poorly in the chemical testing still passed the taste test—and vice versa! One can see that a good deal of thought went into how to obfuscate this fact from the report by leaving out key details. In the end what we are left with is a public that has been made to fear something that's unfounded— the Intentional mixing of olive oil with other types of oil that was never demonstrated. Only alluded to. I'm not saying it can't happen. I'm just saying that nothing in that report showed it.

Some online versions of this story attempt to tie in an unrelated event in New York where marshalls seized 10,000 cases of soy oil that was bottled to look like imported extra-virgin olive oil. It is unclear if that story even happened, and if it did, who the culprit was. Quite possibly an unrelated crime syndicate in New York.

In an multibillion dollar industry, one way to attract customers for your product is to tell everyone that the "other guy" is evil. Still, a few facts were undeniable in the study: Yes, the cheapest brands are the worst quality. As is usually the case, you more or less get what you pay for. Yes, extra-virgin olive oils are easily damaged by heat and prolonged storage. If you are in California, then California olive oil is likely to be fresher. If you are far away from California then there's no clear advantage. In both instances there is no way of knowing how long the bottle sat in a hot warehouse, or just on the store shelf before you picked it up. Shopping at a busy store with an affluent clientele offers a statistical advantage. If you are buying expensive imported olive oil in a small rural store, it probably sat there for a very long time. That shouldn't come as a surprise.

People often associate a greenish color with good taste. Some manufacturers add chlorophyll (from grass) to make it greener — but the label should state that it contains "natural coloring" if you read it. The greenish color of the natural oil is largely proportional to the polyphenols present. These are the molecules responsible for those delicate fruity flavors and floral aromas that you are paying dearly for in the finest products. Among these are terpenes that are very pale yellow and invisible to the human eye in olive oil. So, if we are judging quality by taste and aroma, then an olive oil can be fairly pale in color and still be delicious. If we are judging an extra-virgin olive oil for exhibiting the full range of characteristics that the best possible oils are prized for, then it *must* be green in color — but you don't need the best olive oil for most applications.

Finally, be aware that many olive oils that say they are "Italian" are not made with Italian olives. It has became common practice for manufacturers to ship olives to Italy for processing and packaging so that they can declare "product of Italy" on the label. If you want actual Italian olive oil, then you have to look for (and pay for) the DOP certification (*Denominazione di Origine Protetta*, or Protected Designation of Origin). This ensures it was entirely produced in the stated region from its raw materials through the final production and bottling. If a bottle does not have this designation, you should automatically be suspicious because every Italian manufacturer wants to declare DOP if they can. The cost of inspection and certification is almost nothing compared to the increased market value of the product, so you can take an olive oil "made in Italy" but without "DOP" as a signed confession of this scheme.

PROTEINS

All proteins are made up of chains of building blocks called amino acids. There are 20 different amino acids needed for human beings to live (see picture below). Our body can make 11 of them and the other 9 must be consumed in our diet. These are called the *essential amino acids*, meaning we have to eat them to live. We do this by eating protein. Enzymes in our digestive system tear the protein apart to its individual amino acid constituents so that we can use them for our own body's construction and repair projects.

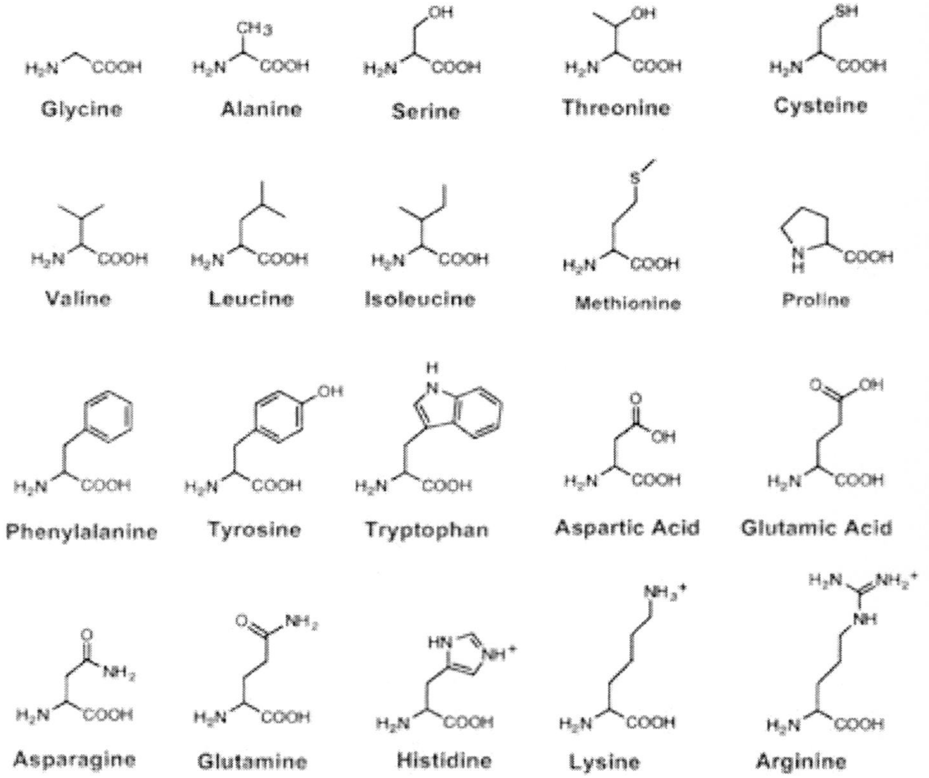

Glycine   Alanine   Serine   Threonine   Cysteine

Valine   Leucine   Isoleucine   Methionine   Proline

Phenylalanine   Tyrosine   Tryptophan   Aspartic Acid   Glutamic Acid

Asparagine   Glutamine   Histidine   Lysine   Arginine

At first this looks very complicated, but you'll notice that they all have the same root structure, as shown in the following diagram:

---

* The nine essential amino acids are: histidine, leucine, isoleucine, lysine, methionine, phenylalanine, threonine, tryptophan, and valine.

Amino Acid Structure

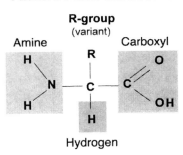

R-group
(variant)

Amine

Carboxyl

Hydrogen

All amino acids have two functional groups attached to a central carbon atom (a carboxylic acid and an amine). The R group is what determines which amino acid it is (*e.g.* for Glycine R=H, for Alanine, R=CH_3). When the carboxylic acid of one amino acid reacts with the amine group of another, they react to form an amide that links them together. This is called a *peptide bond*.

Proteins are Chains of Amino Acids

You can think of these like boxcars on a train. Any number can be put together, and sometimes chains can be hundreds of thousands of amino acids long, but there are always two ends, as shown in the diagram here:

$$^+H_3N - C_\alpha - C - N - C_\alpha - C - N - C_\alpha - C - N - C_\alpha - C - N - C_\alpha - C - O^-$$

**Amino end
(N-terminus)**

**Carboxyl end
(C-terminus)**

Short chains (less than about 20 amino acids) are called peptides, but the principle is the same no matter how long the chain is. In the example of a peptide above there are 5 amino acids.

ADVANCED TOPIC

THREE-DIMENSIONAL SHAPE OF PROTEINS

So far I have kept everything two-dimensional for the sake of simplicity. In the real world, molecules are three-dimensional and proteins have especially complicated shapes. The shape is determined by three different factors...

Hydrophobic / Hydrophilic Interactions

Each region of specific amino acids (R- groups) will cause the protein to curve so that it either faces outward or inward. Some R-functional groups are repelled by water (as in oil and water). These are called *nonpolar*. If there are several like that in a row, they will coil up together away from the surrounding water. On the other hand, some R- functional groups are attracted to water. These are called *polar* molecules. So they will face outward to be in contact with the surrounding water, since bodily fluids are mostly water.

Metal Ions

Some proteins have a kind of pocket (a porphyrin) that holds a metal ion. Two examples of this are the iron in hemoglobin and the magnesium in chlorophyll (as shown on page 60). The metal ion is involved in the funcationality of the protein, such as blood carrying oxygen, which is held in place by the iron atoms of hemoglobin.

Cross Linking

In addition to the two functional groups that all amino acids have, some R- functional groups have other functional groups (see the chart of functional groups on page 47) that attracts them to other functional group along the same protein. In some cases, actual reactions take place to form permanent bridges, especially between sulfur atoms. These tie the protein into a kind of molecular knot at those points.

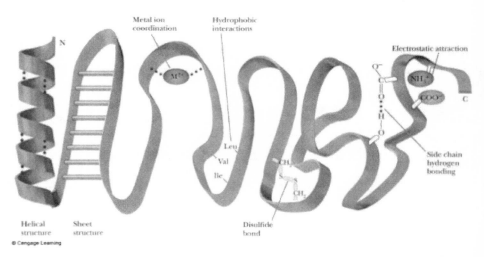

Note: the illustration above is just an example showing how these factors determine the shape. This is not an actual specific protein.

Denaturing of Proteins

The shape of most proteins is critical to their functionality, especially in the case of enzymes that can perform reactions efficiently at body temperature with nearly perfect selectivity (yield). Because they are in a mostly aqueous medium of cell fluids and there is distance between them (at least in terms of molecular size space), proteins are able to maintain extremely complex shapes, such as that of myoglobin shown in the diagram below. This is what makes meat red. It is related to hemoglobin, but it isn't blood.

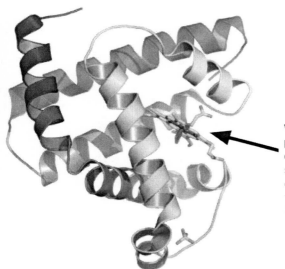

Myoglobin Protein

While hemoglobin contains four porphyrin molecules that each hold one atom of iron. myoglobin has a single such group here. It binds oxygen better, though. This is how tissues get the oxygen from red blood cells.

You can see that such a structure is rather delicate. The exact shape depends on the environment surrounding it being in a narrow range of temperature, a narrow range of pH, surrounded by mostly water that does not contain too many ions such as sodium chloride (salt), *etc.* When you cook something, or add vinegar or citrus (acid) to something, such as in Ceviche, they become **denatured**. That is, they change shape, and it is irreversible. Not only do they get tangled up with each other, but they can cross link into polymeric structures (as in cooked eggs).

Certain chemical solvents and detergents are also methods of denaturing proteins, but those are not involved in cooking.

EMUSIFIERS

The hydrophobic and hydrophilic interactions explained at the end of the previous chapter are due to *polar* and *nonpolar* functional groups (pages 65-66). To better understand this, let's think of polar molecules as being like magnets. They want to stick together. Imagine a big container of magnets and wine bottle corks. If you shake the container for a while, all of the magnets will sink to the bottom and be stuck to each other, while the lighter corks will all be above them on top, right? This is how oil and water separate into layers. Actually it is a bit more complicated in the quantum world of molecules, but the analogy is essentially correct.

Some long molecules are polar on one end and nonpolar on the other. These are called *emulsifiers*. Detergents are all like this. That's how it can make nonpolar oils and fats dissolve in water (which is polar). If you look at the ingredients on almost any bottle of hair shampoo, you'll see one of the first ingredients is Sodium Lauryl Sulfate, also known as Sodium Dodecyl Sulfate (or SDS).

NONPOLAR TAIL   WATER SOLUBLE HEAD

Sodium Lauryl Sulfate

The dual affinity of these molecules acts like kind of a molecular rope tieing the two together.

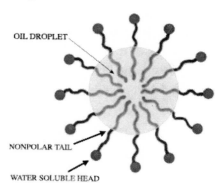

OIL DROPLET

NONPOLAR TAIL

WATER SOLUBLE HEAD

What Is An Emulsion?

Emulsion? =
- Hollandaise
- Mayonnaise & Aioli
- Milk & Cream

Water Fat

- Vinaigrettes
- Butter

Fat Water

Combining two liquids that maintain their distinct characteristics

In Contrast →

Alcohol & Water Solution
- Beer
- Wine
- Cocktails

Water and alcohol can never form an emulsions because they freely mix together.

Common Culinary Emulsions
- Milk
- Vinaigrettes
- Pan Sauces
- Butter

Aioli ↑ Mayonnaise ↑ Hollandaise ↑ Egg Based   Cream

Common Non-Culinary Emulsions
- Floor & Furniture Wax
- Cosmetic Creams
- Some Paints
- Crude Oil
- Asphalt

Soy is one of the most common ingredients in commercially manufactured foods. The protein can be used as a cheap way to manipulate the nutritional information numbers to make a product look healthier. The other reason is that soy is a powerful emulsifier. So are egg yolks, but soy lecithin retains its emulsification property even on cooking. Non-GMO soy flour is available, by the way. One precaution: *There are substantial differences in the emulsifying properties of soy flour from source to source (different brands).

PROFESSIONAL OMELETTE SECRET

For two omelettes: Combine 3 whole eggs, 1 egg white and a teaspoon of soy flour in a bowl. Whisk for a full minute, then let it stand for 5 to 10 minutes. Heat two nonstick skillets with butter on a medium setting (#6 out of 10). When the butter has foamed up and the temperature is about 140°C (280°F), whisk the eggs again briefly and then divide the mixture between the two pans. Do not stir. When the bottom has solidified, add cheese or other desired ingredients (if any), and sprinkle with sea salt. When a skin is almost formed on top, fold the omelettes into halves or thirds, as desired. Cook for another minute or two before turning out.

---

* W.A. Plahar, "Emulsifying Properties of Full-fat Soy Flour", *Ghana Journal of Agricultural Science*, **1977**, (10) pg. 79-83

BROWNING OF FOODS

There are two broad categories of browning foods. Namely enzymatic and non-enzymatic. Examples of enzymatic browning include apples turning dark after they were cut, or how wine eventually turns brown with age. These are rarely useful reactions in cooking. The other type of browning (non-enzymatic) always involves heat. There are three types...

TYPE	TEMPERATURE	WHAT REACTS AND HOW?
Caramelization	Sucrose at 160°C (320°F) but other sugars vary.	Sugar molecules fuse with each other to form a myriad of random highly branched chains.
Maillard Reaction	140-165°C (285-330°F)	Proteins. Sugars react with amino acids, creating many different flavor molecules depending on the specific amino acids involved.
Pyrolysis	170°C (340°F) and above, with the rate increasing rapidly with temperature	All foods react this way as they burn, eventually becoming charcoal. Some amount of pyrolysis occurs in both caramelization and the Maillard Reaction (see text below).

These reactions are incredibly complex, particularly the Maillard reaction because there are as many as 20 different amino acids and five different types of sugars in any given food (more about the different types of sugars in Volume 4 of this series). Even in the case of the simple sugar glucose with the simplest amino acid, glycine (shown on page 64) there are 24 different compounds produced! Recall from the earlier chapter, *Color in Food*, that complex mixtures are usually brown—and that's why foods brown.

The Color Change of Myoglobin

When you cook red meat, the color turns to a grayish brown almost immediately, but that's <u>not</u> the Maillard Reaction at work. Remember that the red color of meat is due to myoglobin, which is red for the same reason that blood is red (but it is not blood). When you apply heat to it, you drive off the oxygen from the myoglobin and also begin denaturing proteins, so it reverts to the color you would expect to see in a complex mixture of organic compounds — brown (as previously explained). The Maillard Reaction doesn't begin until well after this point, and that brown is much darker — and aromatic (myoglobin turning brown has little or no aroma).

Factors Influencing the Maillard Reaction

The presence of fat and salts influence both the rate and the flavors produced by the Maillard reaction. Browning is also influenced by the pH, with both acidic and alkaline conditions increasing the rate. This is why adding baking soda to onions causes them to brown much more rapidly, however the result is not the same as slow cooking because pyrolysis reactions did not have time to take place at the temperature used to cook the onion.

In commercial manufacturing, the Maillard Reaction is often manipulated with the addition of chemicals that home cooks and even most chefs know nothing about. For example, a catalytic amount of a phosphate buffer will greatly increase the rate in most cases and lead to a different balance of flavor and aroma. Because the chemistry is hopelessly complicated to predict, these processes are determined by exhaustive trial and error experiments and generally kept as proprietary secrets by food manufacturers.

Thermolysis and Pyrolysis

Although we tend to think of pyrolysis as simply burning food in a destructive way, before food turns into charcoal there is a type of reaction that influences both caramelization and the Maillard Reaction: *Thermolysis.* This is when chemical bonds are broken by heat. This happens to both starches, which break down to sugars (why fried potatoes get brown), and proteins that break down to amino acids (known as *thermal proteolysis*). Just as browning is influenced by pH, salt concentration and more, so is thermolysis.

Remember that the Maillard Reaction is between amino acids and sugars — not proteins and sugars. So when the temperature reaches a sufficient point for proteins to start breaking down to amino acids, browning suddenly commences. Of course if you keep increasing the temperature beyond that point you soon get to charring. As you can see from the chart on the previous page, the temperature margin between rapid browning and burnt black are not very far apart. Also remember that protein isn't always meat. Flour is often very important in browning, contributing both sugars (thermolysis of carbohydrates) and amino acids to the reaction (more on page 89).

The pH also affects browning. Sometimes dramatically, and not just baking soda with onions, either. The acidity of tomatoes, citrus and even wine can change the outcome.

The final layer of complexity in this ocean of chemical reactions is the physically sticky nature of some of the products of browning. These trap small flavor molecules that otherwise would have been blown off in evaporation, especially if the temperature is fairly low.

BROWN BUTTER

Unlike other fats used in cooking, butter contains a substantial amount of protein (casein and whey) as well as lactose (a sugar). On prolonged heating, butter participates in both caramelization and the Maillard Reaction. That's why *nothing browns like butter*.

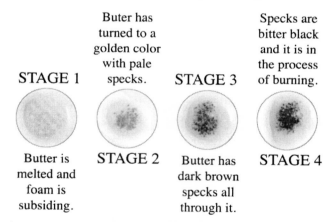

STAGE 1

Butter is melted and foam is subsiding.

Buter has turned to a golden color with pale specks.

STAGE 2

STAGE 3

Butter has dark brown specks all through it.

Specks are bitter black and it is in the process of burning.

STAGE 4

The only real trick is that the heat must be controlled and the time watched, or you will crash through Stage 3 *beurre noisette* (a wonderful hazelnut like aroma and taste) to Stage 4. *beurre noir* (a bitter blackened butter that is generally unpleasant, though actually used in a few old classic French fish recipes.)

Professional Tip: Add a little skim milk powder to the browning butter for an enhanced effect. More brown flecks means more taste.

THE SCIENCE OF ONION FLAVOR

Although there are hundreds of different varieties of onions, they can be grouped into two categories. Namely, mild spring onions, and pungent storage onions. Garlic is in the same allium group as onions, but for now we'll ignore garlic.

Eight Categories of Onions in Cooking

Although there are hundreds of types of onions, technically speaking, only eight account for virtually all culinary appplications.

CHIVES

SCALLIONS / SPRING ONIONS

LEEKS

SHALLOTS

YELLOW ONIONS

WHITE ONIONS

RED ONIONS

SWEET ONIONS

Pungency

By far the most dominant flavor of onions is their pungency. It used to be thought that only one chemical was responnsible for the powerful aroma and sharp taste of onions. Now we know that there are several related chemicals responsible. They arise from the reaction of enzymes in the onion with several **precursor molecules** that all originate from the amino acid, *cysteine*. The structure of this amino acid is shown at the top right corner in the chart on page 64. Understanding the exact chemistry is not important for you, but what you should know is that the reactions that take place happen only after an onion is sliced. This is why whole undamaged onions have almost no odor.

The cell constituents that spill out react with enzymes and the oxygen in the air to produce the compounds we associate with the aroma and mild burning taste of onions. The same is true of garlic, only a different precursor molecule is involved.

A very sharp knife that cuts an onion along the axis (from the root-end to the base) will produce pieces that are the sweetest and least pungent. At the other extreme, grating an onion or puréeing it in a food processor will result in the maximum amount of eye-watering aroma. Onions cook differently depending on how you cut -them, and as such you can't use a food processor on onions in most recipes and achieve the desired result. Unless the recipe specifically tells you to grate or purée them, don't.

Scallions and Spring Onions

Scallions are also known as green onions, and the designations are the most confusing among all of the onions. Several different species are sold under this umbrella term in different countries and by different vendors. The two most common are *Allium fistulosum*, properly called the Welsh onion (even though they come from China!) and *Allium cepa var. cepa,* which is a regular bulb onioin, only harvested when it is still young.

Although botanists might argue otherwise, chefs consider the visible difference to be the bulb on spring onions. Scallions are cylindrical all the way down to the root and have a milder flavor.

Cooked too long at a low temperature, both of these take on a flavor that's soapy or otherwise unpleasant. Therefore they are generally used raw, or cooked quickly at a high temperature.

Chives

First, chives are not scallions. They are the separate species, *Allium schoenoprasum.* Substituting scallions for chives is like substituting spinach for basil. It saves money and looks similar, but lacks all of the herbal character that it is supposed to have. This is an unscrupulous practice often committed by chain restaurants.

Chives are also unrelated to garlic chives or Chinese chives, which are two different names for the species, *Allium tuberosum.*

Their cultivation and use in cooking dates back thousands of years. Chives are a key ingredient in both French and Swedish cuisine. They are also essential in many traditional sausages.

The invention of freeze drying increased the popularity of chives because they retain almost all of their natural flavor, unlike most other herbs. Sprinkled on warm foods, they rehydrate quickly.

Leeks

Botanically, *Allium ampeloprasum.* Looking like a giant scallion, the flavor is the mildest of all the onions. In fact, leeks are served as a vegetable in some parts of Europe. However, like other members of the onion family, leeks have their own flavor characteristics and can't be replaced by any other type of onion if you want good results.

Generally the white part is all that is considered edible. The tough green parts are sometimes used to tie up herbs and spices as a *bouquet garni* in classic French cuisine.

Shallots

There are two different species we call shallots. *Allium oschaninii* is the French eschalot, or griselle. This species is the original true shallot. However, *Allium ascalonicum* is what we normally think of as shallots these days. They look like elongated onions and the skin is colored copper, reddish or sometimes even gray. In south India, a type of shallot that is rounder and more reddish is used almost exclusively and usually just called "onions" in cooking parlance. It is important to keep in mind that throughout south Asia, the term "onion" often refers to what we call shallots. Especially when you see the term "small onions".

Shallots are often described as being like a sweet onion with a hint of garlic, but this is an oversimplification due mostly to the limitations of language in describing the subtleties of flavor. The description also ignores the changes that take place with cooking.

Yellow, White and Red Onions

These are also known as bulb onions. Even though we group these into three color categories, the fact is that they are all variations of the same species, *Allium cepa*.

Aside from their different colors, some are classified as *storage onions*. As the name implies, these can be stored for a long time after they are harvested. The onions you buy in the middle of winter are almost always this type. The same qualities that enables these cultivars to keep from rotting during storage also affects how they cook, and to some extent, how they taste. Furthermore, even though they haven't rotted yet, the longer they are stored, the more changes are taking place. Storage onions are always more pungent. The bottom line is that we don't usually think of onions as having a season, but we really should.

Sweet Onions

These are also members of *Allium cepa*, but in culinary use they are distinct enough to warrant a separate category. As you would expect from the name, these onions have more sugar in them. Not only is the amount of sugar variable in onions, but also the balance between glucose, fructose and sucrose. Because fructose is much sweeter to our taste than glucose, onions like the Vidalia from Georgia, that have a high percentage of fructose, seem to taste sweeter. Vidalia onions also happen to have the least amount of polyphenols, which normally counteract the sweetness of onions. The soil that Vidalia onions grow in is low in sulfur, which limits the production of the sulferous amino acids, further reducing their pungency. Plus, Vidalia onions have been genetically selected for low pungency, so even if you grew them in sulfur-rich soil, they still wouldn't be as strong in taste as most onions. Put these factors together and you can see why sweet onions belong in their own category. Maui, Walla-Walla, Pecos and Bermuda onions are other examples. These are best sutited for salads and fresh applications. Don't use them in cooking unless a recipe specifically calls for them, or the result will be weirdly sweet.

POLYPHENOLS

Polyphenols include tannins (page 60) and flavonoids. The flavor of polyphenols is typicallty astringent (as in tannic acid) or bitter, but some are difficult to categorize and there are thousands of different types. When you hear "bitter and astringent", it may sound unpleasant, but a small amount of such flavor is what we perceive as complexity and, for lack of a better term, quality. Shallots have the highest concentration of polyphenols, being about six times higher than Vidalia onions.

FLAVONOIDS

This is an important class of polyphenol that contributes secondary characteristic flavor notes. Yellow onions have eleven times the amount of flavonoids that white onions do. This gives yellow onions more character, and is why the simpler taste of white onion is sometimes preferable for raw applications where a simple onion taste is desired, as on hamburgers and in Mexican salsa. Flavonoids can bind peptides together to make substances that we perceive as meat-like. This is why yellow onions that are cooked for a long time acquire a meaty taste, and also one way they influence the taste of other foods they are cooked with. For more about flavonoids, see page 85.

ANTHROCYANINS

This is a type of flavonoid that is responsible for the color of a great many plants, including red onions. Anthrocyanins themselves are only in trace amounts and generally don't contribute directly to the flavor. However, they do act as an indicator for where the richest concentration of other flavenoids are, so you can visibly see that they are mostly in the red onion's outer layers. Remember that flavonoids are an important factor in the characteristic taste of similar plants, such as types of onions. The innermost layers of a red onion taste almost the same as a yellow onion because this region contains lower concentrations of flavonoids. Sunlight during growth stimulates flavonoid production, which is why the outer skin layer of onions, tomatoes, grapes and other plants are where the most concentrated color is (anthrocyanin), as well as where the most flavonoids are. Tomato skins are bitter for this reason, but that's another topic that starts on page 84.

Polyphenols as Antioxidants and Health

For many years the idea that eating antioxidants (primarily polyphenols) may ward off cancer, or even cure it. No scientific evidence has been found to support such claims, and recent studies have found that the test tube antioxidant properties of polyphenols are almost immediately destroyed on contact with our digestive system. Polyphenols are essential to foods tasting natural, though. Studies conducted on how cooking affects polyphenols found that steaming preserves the largest amount and microwave cooking is the worst. See Volume 2, page 46, for more about how microwaves are unlike any other cooking method and why this is important to your health.

Problem with Selective Breeding

Only recently has it became known that the polyphenols and flavonoid compounds in onions and some other plants are very beneficial to human health (not as antioxidants). The problem is that these compounds are astringent, bitter and sometimes acrid. The plants commercially grown today have been selectively bred to be lower in these compounds than they were naturally a hundred years ago or more. This isn't GMO technology. It is just selective plant breeding—the same type of breeding that turned wolves into poodles over generations. The public is now used to the watered down and sweeter taste, so reintroducing onions that are as bitter and astringent as they were long ago is probably not going to happen. More important to consider as a thinking chef is that recipes from long ago were using much stronger tasting onions than those today—as well as many other vegetables. This is one reason why recipes that were popular long ago now seem bland. You often have to help them out by adding flavor elements that don't exist in vegetables naturally any more.

ADVANCED TOPIC

CHEMICAL DIFFERENCES BETWEEN TYPES OF ONIONS

There are four important sulfur compounds in onions and garlic that react with enzymes in the cells when they are cut. This is what produces the pungent aroma and taste we associate with the allium family. The precursors are all homologs of Cysteine Sulfoxide.

Cysteine Sulfoxide
(same in all cases)

S-Methycysteine Sulfoxide

S-Propylcysteine Sulfoxide

S-Propenylcysteine Sulfoxide

S-Allylcysteine Sulfoxide

The first of these is S-Methylcysteine Sulfoxide. The "S" in the name indicates that the methyl group (CH_3) is attached to the sulfur (S) atom.

S-Methylcysteine Sulfoxide is also present in vegetables of the *Brassica* family (cabbage, cauliflower, broccoli and Brussels sprouts). The concentration in Brussels sprouts is especially high, but they don't smell or taste like garlic because they do not have the same converting enzymes that are in onions and garlic.

Although only approximate, this chart still helps visualize some of the chemical differences between various types of onions and garlic.

Species	S-Methyl	S-Propyl	S-Propenyl	S-Allyl
Regular Onions	+	++	+++	−
Scallions	+	++	++	−
Leeks	++	++	+	−
Shallots	++	++	+	−
Chives	+	+	++	−
Garlic	++	+	−	+++
Garlic Chives	+++	++	+	+++

In this chart you can see that leeks and shallots have the same relative proportions of these four compounds, yet they are very different in overall taste. This is because the only thing that this chart is expressing are the chemical precursors, and leeks have less of the enzyme to do the conversion. Provide the enzyme, and the reaction takes place on other Cystene Sulfoxides.

This is the chemistry going on in two of the recipes in this book. Namely, *Angel Hair with Garlic and White Wine* (page 208) and *Penne in Arrabbiata Sauce* (page 210). The enzymes in the freshly chopped garlic act on the Cysteine Sulfoxides in both the garlic and the leeks (or onions) that they have been chopped up with. This produces a flavor that no one can reproduce without knowing the secret. This is also why you don't taste the onion in the Arrabbiata sauce. You can use this trick with other members of the onion family to create many subtle unexpected flavors.

CARAMELIZING ONIONS

(CONTINUED FROM VOLUME 2, PAGE 217)

Baking Soda Onions

What it is difficult for non-chemists to wrap their brain around is that just about everything you do in cooking will produce multiple effects. When you add baking soda, you speed up browning and also influence other reactions—and you retard other reactions and prevent *those* flavors from developing. You can clearly taste the difference between onions caramelized with and without baking soda. With baking soda they are sweeter, but there are other different flavors going on in the background, too. Both types of cooked onions are useful, but they are not interchangeable if you have a sensitive palate. For example, I would never make a French Onion Soup with "baking soda onions". For caramelized onions as a component in meatloaf, the sweetness of the baking soda product might be good and it will be faster to cook, to boot.

The same thing goes for temperature. Some people use a pressure cooker to caramelize onions faster. Again, you are promoting other unintended reactions. It might be okay, but the result is not the same as a slow caramelization.

> A general rule of thumb is that *the maximum flavor will be obtained in its purest form when cooked slowly at the lowest temperature possible*.

This is why I braise the onions for French Onion Soup (see video). As Thomas Keller agrees, cooking onions very slowly is the only way to develop their full flavor.

However, that does not mean that braising is the optimum method for cooking <u>all</u> foods. Aside from temperature, time is also a factor. There are always competing reactions in complex mixtures, and an elevated temperature for an extended period of time may cause unwanted side reactions. In some instances, a high temperature for a short time will minimize unwanted reactions. Cooking is always a balance between time and temperature, and even with a good knowledge of chemistry, the optimum combination of time and temperature can only be arrived at by experimentation. This is also the case in organic lab reactions. Pilot studies are carried out with the same materials at a wide range of times and temperatures to discover what gives the highest yield and the least amount of unwanted by-products.

ADVANCED TOPIC

ACTIVATION ENERGY AND REACTION PATHWAYS

The reason you have probably never heard of this before is because attempting to explain it in simple language is almost impossible. There's a reason why it takes years of intense study to get a degree in chemistry. Never the less, I will try to make this concept accessible. To do this I will make some drastic simplifications and avoid working with actual numbers or calculus. We will view everything in relative terms. The very first thing you must understand is that when we measure temperature, it is an average. Within a glass of water (or any other liquid) at room temperature there are individual molecules that are barely moving (low kinetic energy—near what we call freezing) and other molecules that are moving rapidly (high kinetic energy—near what we call very hot). This is quite counterintuitive when you look at that glass of liquid that is neither being heated or cooled. What we call "temperature" is a measurement of the <u>average</u> motion of molecules in a medium. This statistical average is described by the Boltzmann Distribution Curve, as shown below.

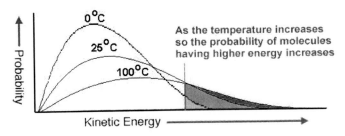

What's important to keep in mind is that reactions can take place even in the refrigerator, though at a much, much slower rate than at a boil. In order for a reaction to occur, the molecules must collide with each other with sufficient energy to overcome a threshold that is called the **activation energy**. You can think of this like trying to dent a car door with your fist. Unless you are quite strong, the indentation will pop back out again most of the time. The analogy to this is *Reaction Pathway #1* in the diagram below.

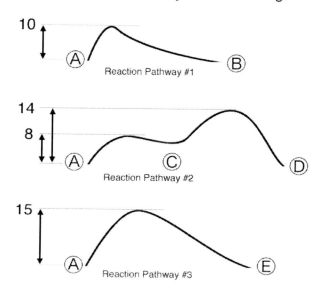

Unless you can apply a force of "10" or more, it will slide back down the first part of the curve to its original shape. If you hit it with a hammer at a force of, say, 20, then a permanent dent will arise, irreversibly transforming Ⓐ into Ⓑ.

Now let's get back to molecules. In a solution containing many different potential reactants. Every molecule of Ⓐ is capable of participating in several different reactions, as shown in these three reaction pathways. Bear in mind that the addition of a catalyst or a change in the pH will alter these curves in ways that we cannot predict. They can only be determined by experimentation.

In *Reaction Pathway #2*, we see the diagram for a different scenario. There is a reaction leading from Ⓐ to Ⓒ that is more favorable than *Reaction Pathway #1* (a relative activation energy of 8 instead of 10), however it is not a one-way reaction because the energy of Ⓒ is still well above zero. In time, some of Ⓒ will revert to Ⓐ because it is not very stable. The lower the energy, the more stable—the less it will be inclined to react. An equilibrium will be reached in which there is some Ⓐ and some Ⓒ in the solution. As long as enough heat is not applied to drive Ⓒ over the next hump to the irreversible state of Ⓓ (stable = zero energy) the solution will continue to be populated by both Ⓐ and Ⓒ molecules. Except that *Reaction Pathway #1* is an option in this scenario, so eventually species Ⓒ will disappear because all of Ⓐ will have been converted to Ⓑ. That is, if you cook it long enough. If you stop it short, you will have remnants of the unwanted product Ⓒ in the dish. You can think of this in the case of long braised dishes that are rushed by impatience.

Now we will introduce another possibility: *Reaction Pathway #3.* This has the highest barrier to overcome (15 instead of 10) so the product, Ⓔ, is easily minimized by cooking at a lower temperature. If Ⓔ was what we <u>wanted</u>, then a high heat would produce the maximum amount, but some Ⓑ and Ⓓ would be unavoiable.

You have to keep in mind that these are all expressed in terms of statistical probabilities. The passing of time is like rolling the dice over and over again. When the temperature is optimal, Ⓐ is being converted to Ⓑ at a steady rate much faster than it is to Ⓓ or Ⓔ because the activation energy for the first pathway is only 10, but it is 14 and 15 for the other two paths, respectively.

If the temperature is too low even for Ⓑ, then entirely different reactions are bound to happen. Meat left on the counter will never cook, but it will certainly rot from enzymes and bacterial reactions that can take place at room temperature.

This explanation is a serious simplification. Activation energy diagrams are really three-dimensional and I've ignored the slopes of the lines. In the real world the diagram also reflects the probability of specific molecules colliding with each other at the right position and angle, which varies with the reaction.

If you have followed this (reading it several times might be necessary), your eyes are now open as to why the selection of time and temperature at each phase in cooking can make the difference between mediocre and superb results in the final flavor. You are an orchestra conductor deciding which instruments play loud or soft.

▼

TOMATO CHEMISTRY

PART 3: A continuation of the topic from Volumes 1 and 2

One of the most challenging aspects of food chemistry is that the compounds that exist in the food we put in our mouths are quickly modified further by chewing and the enzymes in our saliva. What we perceive in our nose, where most flavor is actually detected, is often remarkably different from what it started out as on the dinner plate. Food scientists use the term "nosespace" to refer to these modified volatile compounds (*Taylor, 1994). Some of the flavors we think of as being part of foods, don't exist in that food until they are processed by chewing and saliva. To make matters even more complicated, there are some genetic differences in saliva enzymes from person to person. One such gene is the reason why some people taste cilantro as soapy. Another gene has been found to explain why some people hate the taste of raw tomato, even though most in that same group like cooked tomato.

The chemical difference between fresh and cooked tomatoes is so extreme that it is only through experience we can recognize it's the same plant. Here is a list of the ***primary*** flavor molecules:

FRESH TOMATO	TOMATO PASTE
(Z)-3-HEXENAL	DIMETHYL SULFIDE
(Z)-3-HEXENOL	3-METHYLBUTYRIC ACID
HEXANAL	EUGENOL
1-PENTEN-3-ONE	1-NITRO-2-PHENYLETHANE
3-METHYLBUTANAL`	METHIONAL
(E)-2-HEXENAL	3-METHYLBUTANAL
6-METHYL-5-HEPTEN-2-ONE	BETA-DAMASCENONE
METHYL SALICYLATE	
2-ISOBUTYLTHIAZOLE	
BETA-IONONE	

Footnote: Shortly after this research was published, a patent was filed:

US Patent #5064673 A for COOKED TOMATO FLAVOR, is "a mixture containing dimethyl sulfide, beta-damascenone, 3-methylbutanal, and 3-methylbutyric acid in particular proportions is used to impart or enhance the cooked tomato flavor of food products. In a preferred embodiment, the composition also contains 1-nitro-2-phenylethane, eugenol, and methional." (quoted here directly from the patent).

---

* Taylor, A.J., and Linforth, R.S.T., "Methodology for Measuring Volatile Profiles in the Mouth and Nose During Eating", *Trends in Flavor Research*, vol. 35, **1994**, pg. 3-14

MORE ABOUT FLAVONOIDS

The primary flavor compounds of most fruits and vegetables are relatively small molecules, such as the ones in the list on the previous page for tomatoes and tomato paste. The reason is that small molecules are carried in the air better, where we can sense them in our sinus cavity. Larger molecules like polyphenols affect our taste buds and mucous membranes only on direct contact, so they comprise *secondary* flavor sensations. There are well over 5,000 different flavonoids known. There are no words in the language to describe the specific individual taste sensations of these. They are literally indescribable. However, collectively as a group, we instantly recognize the taste. Just as we can't hum a single F-sharp musical note in any meaningful way on its own, we certainly can recognize a tune that contains that note.

The two words we have that apply to many flavonoids are astringent and bitter, but this is a serious oversimplification. The chart below shows the analysis of six key flavonoids in 12 different wine grapes that were all grown in the same region in the same year. Note that flavonoids are further broken down into six

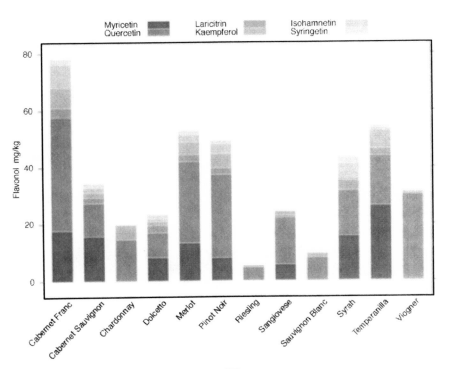

subtypes, one of which is *flavonols*. There is no point in getting bogged down by the nomenclature here. The six flavonols shown on the chart are considered the most important in wine making.

Structure of a Flavonol
The substitutions at R determine which specific flavonol it is.

Example: Quercetin

Curiously, recent research has found that polyphenols can reduce the aroma of some primary flavor molecules, while still being important in the characteristic taste. Balance is the key.

Generally speaking, flavonoids are concentrated in the outer skin, with higher concentrations in strongly colored plants. Thus, red grapes have more flavonoids than green grapes, and red onions have more than yellow onions. Knowing this, the graph follows the general trend that we would expect, with the most astringent (tannic) red grapes having the highest percentage of flavonols, and riesling—a sweet, pale grape—having the least. The individual balance between flavonols is part of what gives each grape its unique taste, and variations resulting from soil conditions and the weather during the year result in variability in the concentration of individual flavonoids, and what we collectively term the **character** of the wine.

The same thing is true of other fruits and vegetables, too. It's just that wine is unique for being so widely studied and appreciated in all of its subtle nuances. If there was a similar interest in tomato sauce, we would have every bit as much to quibble over between cultivars, climate, soil effects and specific vintages.

See also: C. Lund, "Effect of Polyphenols on Perception of Key Aroma Compounds", *Australian Journal of Grape and Wine Research*, 2009 (15), pg. 18-26

WHY FRESH TOMATOES ARE USED WITH PASSATA

Flavonoids

The flavonoids are mostly in the *pomace* (the skin and seeds of the tomato), but the pomace is strained off in passata. The pomace is normally sold off separately by tomato processing plants to be dried and mixed in with livestock feed as a source of protein. So most of the tomato character that is expressed by flavonoids is missing from passata. That's okay if tomatoes are not the center of attention in the dish, because the subtle nature of the flavonoids will be overwhelmed by other flavors going on in many instances. But, when you are making tomato paste, or anything where tomatoes are front and center, you need the flavonoid content.

So why not just use fresh tomatoes and forget the passata altogether? Because you can't buy vine ripened tomatoes most of the year, and even when you can they won't compare in quality to the Italian tomatoes that passata was made from, particularly in the case of San Marzano tomatoes. See Volume 1, pages 19-20.

The Maillard Reaction

There are two types of browning reactions, as explained in the previous chapter. Namely, the caramelizing of sugars and the Maillard Reaction, which occurs between proteins and sugars. Tomato pulp contains less than 1% protein and a lot of water. When you cook tomato pulp for a long time, it is the sugars that caramelize. However, the seeds and skin are about 30% and 16% protein respectively, so they can participate in The Maillard Reaction, given the chance. But the Maillard Reaction requires a temperature of 150°C (300°F) minimum, which can't be reached in the mostly aqueous medium of tomato purée. Therefore, the skin and seeds must be cooked first in oil to get a good yield.

If you lightly brown the pomace you get some lovely deep tomato flavors. In the YouTube video, *Real Italian Tomato Paste*, I add in whole tomato skin-side down at the start to get some of that flavor in the sauce, but we can take that further. The recipe for the full version is on page 122.

COMPARISON OF FLOURS

All measurements in the table are per 100 grams (about 3.5 ounces), or roughly a cup in volume when freshly sifted. Note that some manufacturing methods can produce flours with substantially different values from the average nutritional statistics shown here.

FLOUR	CALORIES	FAT	CARBOHYDRATES	PROTEIN
WHITE "ALL PURPOSE"	380	1.0 g	77 g	10 g
WHOLE WHEAT	340	2.5 g	72 g	13 g
HARD RED WINTER WHEAT	330	2.0 g	71 g	13 g
DURHAM (SEMOLINA)	360	1.1 g	73 g	13 g
OAT	420	9.5g	68 g	19 g NO GLUTEN
RYE	350	1.5 g	75 g	11 g LOW GLUTEN
CORN	375	1.4 g	83 g	5.6 g NO GLUTEN
BARLEY	340	1.6 g	75 g	11 g
BUCKWHEAT	340	3.1 g	71 g	13 g NO GLUTEN
GRAM (CHICKPEA)	390	6.7 g	58 g	22 g NO GLUTEN
SOY - DEFATTED	330	1.0 g	38 g	47 g NO GLUTEN
SOY - FULL FAT	436	22.0 g	35 g	35 g NO GLUTEN
PEA (PEASEMEAL)	434	1.2 g	75 g	33 g NO GLUTEN
* AMARANTH	370	7 g and up	65 g	14 g NO GLUTEN
* HEMP FLOUR	400	9 g and up	51 g	28 g NO GLUTEN

* Amaranth and hemp are seeds that are first pressed to extract the oil before they can be ground into a flour. Fat content varies widely depending on how much extraction is done.

TOASTED FLOURS

Toasting flour in a dry pan is common in Cajun cuisine, as well as some older French recipes (which is how it came to Cajun country). This adds a nutty taste that is most often used to thicken soups and stews. In other parts of the world, especially South America and parts of Asia, toasting flour is also traditional, only with other flours than wheat. Curiously, this technique is seldom mentioned in cookbooks and the whole idea is surprising to most people in other parts of the world.

As you can see in the table on the previous page, all flours have carbohydrates and protein. The carbohydrates in flours include starch, fiber and sugars. Therefore toasting flour results in a mixture of caramelization and the Maillard Reaction in various proportions depending on the specific flour.

Flours also contain small amounts of other substances that give each of them their unique flavor. After all, if they literally only contained starch, sugar, protein and a small amount of fat, they would all have the same completely neutral taste. The fact that barley tastes nothing like buckwheat is due to the presence of substances that are unique to each grain. Of all of these, bleached wheat flour is the most neutral tasting—until it is toasted. Just don't go too far and burn it.

Like other fine ground seasonings, toasted flour has a lot of surface area for flavor molecules to react and vaporize, so the sooner you use them after toasting, the more flavor they will have.

A NOVEL APPLICATION

When you cook foods, some of the volatile flavor components escape into the air and some react with other substances, changing their flavor (or more often, losing it completely). We accept this as a normal result of cooking, and we are accustomed to cooked foods tasting different from raw foods. It's so natural that we don't even think about it. For example, the completely different nature of a fresh tomato vs. cooked tomato paste as previously shown on page 34. This shows how wide our tolerance is for detecting "tomatoes".

What if we could trap some of these flavor molecules and store them away so that they would reach our palate instead of being oxidized or blown off in the air? My initial inspiration for this idea was the old British innovation of using turmeric in ovens to absorb curry flavor components (see Volume 2, pages 41-42). The problem is that turmeric has a strong flavor that doesn't play very well in most other cuisines. It can also react and change other flavors.

Then I thought about the technique of column chromatography in organic chemistry labs. This is used as a method of separating components in a mixture based on the different affinity they have for an absorbant (usually silica or aluminum oxide). The separated components can be washed off of the absorbant at any time. Until then, they remain trapped on the vast porous surface area of the powdered material.

We wouldn't enjoy eating silica or aluminum oxide, but we can eat flour. Only it turns out that wheat flour, which is the most neutral, just doesn't function very well as an absorbant. Other flours with more protein and fat work much better. This is the basis for many of the novel spice blends and coatings starting on page 217.

SELECTION OF FLOURS IN SPICE BLENDS

The selection of the specific flour is based on two criteria: Taste and chemistry. For example, amaranth flour was chosen for the *Beautiful World* seasoning blend because the high fat content, is necessary to absorb compounds in the arugula leaves and the bitterness is complimentary. Oat flour wasn't as complimentary.

In some instances, the reverse situation exists, where there are fat-soluble compounds that need to be toned down for a useful product, such as dates with semolina. In the end it comes down to a matter of experimentation to come up with a combination that has broad use.

For Further Reading

"Structural Modification of Wheat Gluten by Dry Heat - Enhanced Enzymatic Hydrolysis", *Food Technology and Biotechnology*, **2012**, 50 (1)

"Thermal Stability of Wheat Gluten", *Cereal Chemistry*, **1980**, 57 (6)

ANIMAL FAT VS. VEGETABLE OIL

As misinformed as many people are about MSG being bad for you, even more people fear animal fat or lard. Ever since our childhood the media has been telling us that the animal fats that humans had been eating since the days of Fred Flintstone are killing us, but somehow these chemically extracted plant fats that have only been around for a short time are beneficial. This is a case of where the ball got rolling and couldn't be stopped. An entire industry was created to manufacture and market vegetable oils for cooking and hydrogenated vegetable oils for commercially manufactured foods. A trillion dollar industry doesn't just go away. A more recent example of this phenomenon is the gluten-free food industry (see page 42). Once such an industry is entrenched, no amount of science is going to unseat it. The fact is that recent research has shown that animal fats are perfectly fine to consume, and may be much healthier than vegetable oils. Certainly they are healthier than hydrogenated soy or cottonseed oils, which are a large part of most commercially manufactured food products.

From a practical standpoint, restaurants are going to continue using vegetable oil for just about everything because it is much less expensive and because it has a higher smoke point than animal fats do—and restaurants usually cook foods at a high temperature to get the job done as fast as possible. Besides, at this point most consumers prefer vegetable oil, and most would avoid a restaurant that bragged they used lard for their cooking. Up until 1990, McDonald's french fries were cooked in beef fat. They changed to vegetable oil to abate the bad press they were receiving, and the fries have never tasted the same since.

The point here is to try and get over whatever fear you might have about animal fats. They can add a lot of flavor and unctuousness to a dish, and as a home cook you don't always have to cook at a super high heat because you have eleven orders backed up and more coming in every minute.

However, there are practical considerations that must be observed. The most obvious one is where you are going to get this animal fat from? While there are a few online sources for small tins of rendered fat at a high price, your best source is your own kitchen as you roast meats and make stocks. Instead of scraping all of it into the trash, you can often strain it and use the fat for cooking something else later. Not always, though. For example, if you were cooking something with chili peppers or some other strong spice blend, then the fat that ran off is going to have those flavor components in it, which likely will clash with the next dish you are trying to make. Stock fat is less of a problem in that regard, but still quite herbal compared to plain rendered fat.

TYPE	SMOKE TEMP	SATURATED	NOTES
Canola	205°C/400°F	7%	Neutral flavor and inexpensive, but recent studies suggest health risks.
Corn	232°C/450°F	13%	Frying foods in GMO corn oil is linked to several types of cancer.
Olive	200°C/392°F	14%	High quality olive oil quickly loses the sweeter fruity notes on heating.
Peanut	225°C/437°F	17%	Retains the peanut flavor after cooking, so best for Asian dishes.
Sunflower	227°C/440°F	10%	Arguably the best general purpose cooking oil for cost and health.
Soybean	238°C/460°F	16%	The type that doesn't turn rancid easily is a Monsanto GMO product.
Coconut	175°C/350°F	90%	Low smoke point. Does the good outweigh the bad? The FDA says no.
Beef Tallow	205°C/400°F	50%	Add richness, but makes foods taste heavy if very much is used,
Chicken Fat	190°C/375°F	30%	Adds chicken taste. Unpleasant mouth feel in large amounts,
Bacon Fat	190°C/375°F	42%	Adds a slightly smokey flavor, whcih can be either good or bad.
Duck Fat	190°C/375°F	10%	The main drawback is the cost (unless you cook a lot of duck).
Butter	175°C/350°F	51%	Burns easily, but also browns foods the best and is not very expensive.
Clarified Butter	245°C/475°F	53%	Very high smoke point, but not very much flavor. See Ghee on next page.

Of all of the vegetable oils, extra-virgin olive oil has the best health record, but it's also the most expensive—but don't waste your money on the finest artisanal product for cooking because the volatile fruity flavors are burned off at high temperatures.

All cooking oils and fats are a mixture of different molecules. Unrefined and unfiltered vegetable oils are reactive because of the alkene groups they contain (see page 47) - which is what unsaturated means. They also have other substances in them that oxidize (burn) easily. Such oils are only suitable for cold dressings and other applications where they will not be heated. Coconut oil is highly controversial. Although you can find seemingly credible enthusiasts, the FDA and the AMA both regard it as very unhealthy.

Animal fats fair much better because they are less reactive, being more saturated (alkanes instead of alkenes). They contain trace amounts of proteins and sugars that can actively participate in browning. Still, the best choice for browning is invariably butter. The only problem with butter is that it burns easily, so browning must be carried out at a lower temperature for a longer time than with oil, or even animal fat. That's a problem if you are trying to brown something on the outside, but leave it rare in the center. People who brown steaks in butter either have bitterness on the surface or they end up with a thick region of grayish meat between the surface and the center, which compromises the overall flavor.

Mixing Butter with Oil

This is commonly believed to increase the smoke point of the butter, but the reality is different. The effect is dilution and not any physical change to the smoking point. That is, the proteins in the butter (milk solids) that will combine and burn are separated by a greater distance, which is actually what slows the reaction down. Substances in the butter than burn on their own still burn.

Clarified Butter and Ghee

Clarified butter is melted butter in which the milk solids (proteins) have been skimmed off. It has a very high smoke point and a more flavor than vegetable oil.

True ghee is slowly cooked butter. The solids are strained out at the end, but not before they have contributed some flavor to the ghee (when made in the traditional method - which is often not the commercial product, mind you). In India it's common to add spices to the butter while it is cooking, and it is still just called ghee.

STOCKS AND BROTHS

These terms are frequently used interchangeably, but they are not the same. The confusion arises mostly because there is a gray area between them. Technically, the emphasis on a stock is the bone gelatine that's dissolved in it. This is meant to give body to sauces and stews, as well as provide a relatively neutral medium for soups and braises, etc. It is made primarily with the bones of the animal (or fish) that the stock is intended to be used with. The exception being veal stock, which is so neutral that it can be used with anything, including fish.

Broth, on the other hand, must contain a large proportion of meat from the animal as well as some bones. The emphasis is on the flesh rather than the gelatine from the bones (though it needs some of that, too). There are white (light) broths and brown (dark) broths, and the difference is that dark broths employ the Maillard Reaction. Of course, you can't completely avoid the Maillard Reaction when cooking anything, but a well made white broth involves a procedure to minimize browning.

Pressure Cookers

Because of the high temperature in a pressure cooker, any meat will undergo the Maillard Reaction as it cooks. Therefore you can automatically forget about pressure cooking light broths. For dark broths and some stocks, a pressure cooker is a great choice providing the bones are small, as with poultry and fish (although brown fish stock is almost never called for). The problem with pressure cooking beef or veal broth is that the bones do not have enough surface area to dissolve before destructive reactions take place. Nothing evaporates, since it is sealed, but flavor molecules oxidize and react with each other. The situation is even worse when trying to make beef or veal stock in a pressure cooker. A good stock is prepared at a simmer of around 80°C (175°F), but a pressure cooker operates at around 120°C (250°F). The necessity for cooking at a minimum temperature was explained back on page 80.

COMPOUND BUTTERS

The most famous compound butter is *Maître d'Hôtel*, which classically incorporates lemon juice, fresh parsley, salt and pepper. There are variations, most frequently being the addition of minced shallots, tarragon and/or chervil. This was originally made fresh and served on top of steak and Chateaubriand. That is, back when people ate more buttered steak and there was still tableside service. An overdressed *Maître D'* (often with gold braid and gleaming medals as if he was an admiral or a South American dictator) would glide up to your table and prepare the butter from a rolling trolley equipped for the task. However, ironically compound butters improve upon being refrigerated for a day or two before use, so it was mostly for show—again, back when such theatrics were part of the experience of a dark—very dark—romantic restaurant where you would take a beloved date, as opposed to a family style restaurant where the lighting is like high noon in the Mojave and the only theatrics are the screaming kids in the next booth over.

Below are three nation-themed compound butters. Try them on steamed vegetables as well as grilled fish and meats. Compound butters can be classed as *tertiary* seasonings (see page 217).

Sicilian: 100 grams butter softened at room temperature, 2 teaspoons olive oil, 1 tablespoon Sicilian Spice Blend (page 226), 1 teaspoon freshly grated orange zest, 1 teaspoon crushed garlic. Mash butter and other ingredients together with a fork. Scrape out onto parchment paper and roll up. Freeze for 1-2 hours, then transfer to refrigerator for use (or leave it in the freezer for long term storage).

Finnish: 100 grams softened butter, 1 tablespoon Finnish Spice Blend (page 232), 1 tablespoon finely minced fresh dill, 2 teaspoons sunflower oil, 1 teaspoons finely minced shallots, 1/2 teaspoon ground white pepper. Otherwise same procedure as above.

French: 100 grams butter, 2 teaspoons olive oil, 1 tablespoon Provençal Spice Blend (page 231), 2 teaspoons finely minced shallots. Once again, follow the same procedure as before.

FLAMBÉE SCIENCE

Speaking of restaurant theatrics back in the day, another bit of drama that's now rarely performed due to insurance underwriters and skittish fire marshalls, is the tableside flambée. That is, to set fire to a dish with a high-proof liquor as the fuel. The potential for igniting guests or the restaurant itself has largely extinguished this practice (pun intended). The question is if it's just an entertaining circus act, or if it actually changes the flavor of the dish itself?

The latter. In three important ways:

1. REDUCTION OF ALCOHOL The most obvious is that the alcohol in the spirit is rapidly burned off before it has much of a chance to mix with the rest of the dish, thus leaving more of the flavor components of the liquor rather than the booze. Actually, a few decades ago it was common practice to only half-flambée something. A cover would be placed over the top to stop the flame from burning off all of the alcohol because customers wanted to taste the hard liquor in sauces and desserts.

2. BROWNING REACTIONS Burning alcohol has some common ground with the propane torch used for making Crème brûlée, only with a more natural and pleasant taste. Sugars are caramelized as well as other reactions on the surface that can only occur at a very high temperature. This is a classic French technique for improving the flavor of farmed champignon mushrooms that you have probably seen me do in videos. The mushrooms are flambéed with cognac. The Maillard Reaction takes place on the surface, as well as some other reactions, undoubtedly. Mushrooms are absorbant by nature, so they soak up the essence of some of the cognac flavor molecules—but only the ones that are not especially volatile, and that's the key. The flavor profile of the liquor itself is altered dramatically by the burning process. This is the reason that bourbon is not as commonly used to flambée foods. Bourbon leaves only a trace of burnt aroma, and sometimes even some unpleasant bitter notes, depending on the food. The liquors of choice for flambéeing are rum, cognac and tequila (however tequila is rarely suited to anything other than Mexican or Tex-Mex dishes).

3. INCINERATION OF VOLATILE COMPOUNDS — A flame creates strong upward convection. Air moves up and away from the surface of the food, carrying volatile components along with it, where they are incinerated in the fire. Whether this is beneficial or not depends on the food and the amount of time the flame lasts, because this takes longer to affect the overall flavor than the other two flambée reactions previously mentioned. Still, the fact is that the flavor profile is changed by this process.

The well known dessert, *Bananas Foster*, benefits from all three of these processes. First, most people these days don't like a strong taste of alcohol in their food, and so it is largely burned off before it has a chance to soak into the bananas. Second, the sugars in the bananas, as well as the sugar in the recipe itself, are caramelized by the flame to produce a deeper flavor. Finally, bananas exude ethylene (a gas for ripening) that has a fairly musky aroma, as well as some volatile flavor components that are grass-like in aroma and taste. Some of these are carried off and burnt in the air during the flambée, which is to the benefit of a dessert.

A flambée is not always a good approach, though. There is no such thing as a one-size-fits-all solution in cooking. Every case is different, as well as opinions about what tastes better.

> *"My cooking is so bad that my kids thought Thanksgiving was to commemorate Pearl Harbor."*
>
> ...Phyllis Diller

WHY THERE'S NO CHINESE FOOD

Chinese restaurants are extremely rare in Russia, and the few that exist are all as empty as they are horrible in my experience. They beg the question if Russians dislike Chinese food because the only kind they've had has been appalling, or if the flavors are just too alien? Finding a qualified Chinese cook in Saint Petersburg is very difficult. I know because one day I was called up by one of the biggest restauranteurs in the city who told me he had been trying to find someone who could cook Chinese food here for nearly two years. He had a building and a desire to open the first quality Chinese restaurant. My first task was to prepare 20 dishes for a table of his Russian business colleagues to sample. However, three of the items were to be non-Chinese, and dictated to me as a steak "with all the trimmings", Olivier salad and a pastrami sandwich. The latter is unheard of here, but the guy was a wealthy world traveler and had loved the pastrami in New York's Stage Deli. This also explained his unusual taste for Chinese food, because he'd had it in the United States instead of in Russia. No problem. American-style Chinese dishes were planned.

Pastrami is not even for sale in Russia, so my next task was to cure the meat myself at home well in advance of the night of the presentation. I also enlisted the aid of a sous chef from a place I'd worked at here, because putting up 20 completely different dishes in an hour (his requirement) in an unfamiliar kitchen is impossible for anyone working alone. Finally the night arrived and everything went off as smoothly as could have been hoped for. Although it seemed mighty strange to send out a pastrami sandwich between Kung Pao Chicken and Shrimp Fried Rice, but that was the sequence he had dictated. I kept getting feedback from the single waiter who had been brought in for this one-table banquet, that the boss was absolutely ectastic, so I had nothing to worry about. My sous chef was especially thrilled at the prospect that he might become the head chef at this new prestigious restaurant downtown. I wanted it to work out for him.

Finally the verdict came back. The boss raved that every dish had been even better than the one before it! Unfortunately his colleagues only loved the steak. The pastrami was unfamiliar to them, and the Olivier wasn't any better than what they had in other high end restaurants. Most Russians who have the money to open restaurants have no taste for exotic foods. They ended up opening it as a bakery.

Russian Recipes

As I have said before, there is no true Russian cuisine. Every "Russian" dish is either a version of a recipe from nearby regions (especially Poland and Ukraine) or something that was introduced by a foreign chef working for Russian nobility long ago. The reason for this is explained in Volume 1. Let's compare the cultural and historical figureheads of France and Russia. Both nations have legendary composers, authors, military leaders, scientists and inventors—but while France has many famous chefs known the world over, Russia does not have even one. Ask a Russian who is a famous Russian chef, and they will have to admit they don't know.

One might argue that there is a valid answer to this question, though: Vladimir Mukhin, of the White Rabbit restaurant in Moscow. The restaurant is rated as one of the 50 best in the world, after all. He is a Russian, but he trained in a French Michelin star restaurant. The problem is that his dishes are either purely French or Russian standards elevated with dollops of foie gras and shavings of truffles. So, while Mukhin is doing a fine job catering to visiting tourists and Russian billionaires, he's unknown to the average Russian. He's in a world of his own. Compare his influence on Russia to that of Paul Bocuse on France, or Marco Pierre White on England, or Julia Child on America. There is no equivalent in Russia, and there never has been.

It's not his fault, though. This is a culture that is about as interested in fine food as Americans are interested in glassblowing. A very tiny audience, indeed. I have studied the old cuisine and created some dishes that can genuinely be called New Russian—recipes that received critical acclaim in the press here—but I'm a visiting foreign chef, too. No Russian chef would invent such things.

VODKA FRIED CHICKEN
WITH "WOLF AND BUNNY" SAUCE

The recipe for the sauce was not included in the video because that would have meant explaining how to make the Novgorod Spice Blend, which has several steps in itself. The sauce gets its name from a popular Russian cartoon.

FOR THE MARINADE

700g (1.5 lbs)	Chicken pieces (wings or small thighs)
70ml (2.5 oz)	Vodka (see note above)
30ml (1 oz)	Apple Cider Vinegar
3-4 cloves	Garlic
1 whole	Red Serrano Chili Pepper
1 whole	Chicken Stock Cube, ideally Knorr

PROCEDURE

1. Scrape the seeds and membranes from the chili. Crumble the stock cube into a stick blender cup, or crush it in a mortar. Add the other ingredients to the cup (except the chicken) and purée.

2. If you are using chicken wings, trim off the wing tips and divide the pieces across the joint into wingettes and drumettes, as they are called.

3. Massage the marinade into the chicken. Refrigerate for 1 to 3 hours.

FOR WOLF AND BUNNY SAUCE

50ml (1.75 oz)	Vegetable Oil
30ml (1 oz)	Vodka
30ml (1 oz)	Ketchup
1 T	Novgorod Seasoning Mix (page 237)
1 1/2 teaspoons	4-Mix Peppercorns (black, white, green, pink)
1 teaspoon	Honey
1/2 teaspoon	MSG
1/4 teaspoon	Cayenne

4. MAKE THE SAUCE. Grind the mixed peppercorns in a mortar. Combine with the other sauce ingredients in a small sauce pan. Whisk together as you bring the mixture to a boil. As soon as it reaches a boil, set it aside to cool. This is what's called a "broken sauce" - it separates into two components that you serve in a small bowl for dipping.

FOR THE COATING

100g (3.5 oz)	Flour
1 T	Black Peppercorns, whole
2 teaspoons	Rosemary, dried
1 1/4 teaspoons	Baking Soda (not baking powder)
1 teaspoon	Oregano, dried
1 teaspoon	Thyme, dried
1 teaspoon	MSG (optional)
1 teaspoon	Sugar
3/4 teaspoon	Salt
1/2 teaspoon	Paprika
1/2 teaspoon	Turmeric
1/2 teaspoon	Parsley, dried

---

5. MAKE THE COATING. Put the black peppercorns and rosemary into an electric spice mill and pulse several times. You don't want to turn it into a powder—you want to coarsely grind it. Now mix this with all of the other ingredients in a large enough plastic box (with a lid) to hold all of the chicken.

6. Get the oil in the deep fryer hot before you proceed. The chicken needs to hit the oil roughly 10 minutes after it gets the coating and your deep fryer will take longer than that to reach 160°C (320°F).

7. Shake the excess marinade from each piece of chicken and add it to the container. When you have all of the pieces in the box, put the lid on and shake it to evenly coat the chicken.

8. Deep fry wing pieces for 5-6 minutes. Thighs will take 11-12 minutes. Allow to cool well before eating.

+

RUSSIAN WINTER PLOV

As I said at the start of this section, there's really no such thing as an authentic traditional Russian dish. This is adapted from Uzbeki cuisine, but the real origin is undoubtedly older than that. This version has more meat it in than would be traditional. As a result, the rice is stickier and heavier than it would normally be. You can make this same dish with half the meat and the same amount of water, rice and vegetables and you'll have a lighter textured plov, if you prefer.

680g (24 oz)	Pork Roast, cubed
250g (8.8 oz)	Rice, long grain (rinsed in cold water)
100g (3.5 oz)	Carrot, grated
100g (3.5 oz)	Onion, chopped
6 cloves	Garlic, chopped
2 1/2 teaspoons	Paprika (sweet)
2 teaspoons	Coriander Seeds
2 teaspoons	Kosher Salt (coarse)
1 1/2 teaspoons	Garlic Powder
1 teaspoon	Black Peppercorns
1/2 teaspoon	Dill, dried
1 whole	Bay Leaf
50g (1.8 oz)	Ketchup
15ml (1 T)	Lemon Juice
30ml (1 oz)	Sunflower Oil, preferably unrefined
For plating: Fresh Dill, Carrot and Red Chili Peppers (optional)	

SUNFLOWER OIL

Unrefined sunflower oil is a traditional ingredient in Russian cooking that is uncommon elsewhere. It is nutty and delicious. Look for it.

PROCEDURE

1. MAKE THE SPICE PASTE. In an electric spice mill grind together

the paprika, coriander seeds, coarse salt, garlic powder, black peppercorns, dried dill and bay leaf. In a large bowl combine the ground spices with the ketchup, lemon juice and the sunflower oil.

2. Tenderize the pork as shown in Volume 2 under *Steak Mastery* . Then cut it into cubes and mix with the previously made marinade. Refrigerate for between 1 and 4 hours.

3. Spread out the pieces of meat on a metal baking tray. Roast at 220°C (450°F) for 10 minutes with fan assist ON. Then turn the pieces over and roast another 10 minutes. Turn them back again for 3-5 more minutes.

4. Remove the meat to a bowl to hold. Deglaze the metal roasting pan with 500ml (2 cups) of hot water. Pass the liquid through a sieve two times and discard the solids. Refrigerate the liquid until the fat at the top is solid.

5. Scrape the fat off of the top. Add the fat to a hot pan and sauté the carrots and onions in the fat on a high heat (#8 out of 10).

6. After 5 minutes add a teaspoon of salt. Stir.

7. After 1-2 minutes, add the rinsed rice. Stir.

8. After another minute add the garlic. Stir.

9. After another minute add the previously browned meat. Stir.

10. After another minute add the liquid from deglazing the pan. Stir.

11. When the mixture comes to a simmer, reduce the heat to low (#3 out of 10). Place an inverted plate on top that almost reaches the edges of the pot over the mixture. Cover this with a lid. Simmer for 30 minutes. The key here is to have the mixture cooking at the right temperature. Too low and it won't have a good deep flavor. Too high and the bottom will burn.

12. During this time, mince some fresh dill and grate a little more carrot. for a garnish. If you are going to use the red chili peppers for garnish, then heat a pat with a little oil in it and roast the peppers to soften them.

13. When the time is up, stir in the carrot and dill before plating. Add the red chili to the plate. A roasted head of garlic is also a common garnish.

TSATSEBELI
(RUSSIAN VERSION)

Often spelled "satsebeli" in English, the word simply means "sauce" in Georgian, which should tell you how popular it is. The taste is kind of a spicy ketchup and it is used on everything.

1 jar	Pickled Tomatoes (see notes below)
200g (7 oz)	Tomatoes, vine ripe
2 T	Sugar
1 1/2 teaspoons	Corn Starch
1 whole	Red Serrano Chili, sliced lengthwise (optional)
1/2 teaspoons	Paprika, or Hungarian Spice Blend (page 229)
1/2 teaspoon	Khmeli Suneli (see note below)
1 teaspoon	Red Wine Vinegar

ADDITIONAL NOTES

Russian pickled tomatoes have a unique flavor that you can't duplicate because they are produced using some regional Russian herbs that are not exported, and not even sold in Russian stores. The size jar you need for this is the one that's around 400 grams (14 ounces) of tomatoes and around 850 grams (30 ounces) total weight (both tomatoes and brine). Pickled tomatoes are also often sold in very large jars of up to four liters, but as a general rule the larger sizes are inferior in quality. Stick to the smaller size jars to help ensure you are getting a quality product.

Khmeli Suneli is a Georgian spice blend. Like the Russian tomatoes, Khmeli Suneli is made with some herbs in Georgia that are not exported individually. Note that you need to replace this every year at the most, because even though it may smell fine, it will taste old.

PROCEDURE

1. Drain the pickled cherry tomatoes through a sieve, but keep the liquid that runs off. If there are any large fibrous pieces of herbs or other plant matter, pick them out and discard. Weigh the drained tomatoes and then add enough of the strained liquid to bring the weight up to 600 grams (21 ounces).

2. Put the tomatoes and marinating liquid into a blender along with the fresh tomatoes, sugar and corn starch. Blend for a full minute.

3. Pass the resulting mixture through the fine grate of a food mill. Put the resulting liquid into a sauce pan and bring to a rapid simmer.

4. Simmer until the volume is reduced by about half. Now add the red chili pepper if you want the spicier version.

5. Continue simmering until the volume has been reduced to the desired thickness, but about 200 grams (7 ounces) is a good stopping point. When it gets to within a few minutes of the end point, add the Khmeli Suneli.

6. Transfer to a jar and stir in the red wine vinegar. Cool to room temperature before moving it to the refrigerator.

In this Georgian cartoon, the pet fish is apprehensive next to a bottle of tsatebeli, a sauce used on just about every food, including fish.

PASTA WITH MUSHROOMS AND BEAR SAUCE

This is an example of the New Russian cuisine, which borrows from other nationalities but breaks a lot of rules.

300g (10.6 oz)	Rotini / Fusilli Pasta
60g (2 oz)	Onion, diced
30g (1 oz)	Celery, peeled and diced
120g (4 oz)	Porcini Mushrooms, frozen
2 t	Bacon Fat (optional)
1-2 T	Parmesan Cheese
60ml (2 oz)	Cream (optional)

FOR THE BEAR SAUCE

120g (4 oz)	Tomatoes, fresh - in large pieces
90g (3 oz)	Onion - in large pieces
1 clove	Garlic
60ml (2 oz)	White Wine, dry
1/2 teaspoon	MSG (optional)
1/2 teaspoon	Salt
1/4 teaspoon	Sugar
2 T	Bear Demi-glace (see note below)

Vegetable Oil, Butter, Fresh Herbs

BEAR DEMI-GLACE

Obviously this is an ingredient that most people will find impossible to obtain, so you can substitute ordinary veal demi-glace, or just leave it out and make the dish vegetarian. Bear meat is hard to get even in Russia.

PROCEDURE

1. MAKE THE SAUCE. Add either a tablespoon of bacon fat or vegetable oil to a hot nonstick pan. Put the large pieces of tomato and onion into the fat. Cover and reduce the heat to medium (#6 out of 10).

2. Turn the piece over after about 8 minutes. Continue cooking another 4-5 minutes until there is a lot of darkening and fond built up on the bottom.

3. Crush the garlic clove against the side of a knife and add it to the pan before deglazing it with the white wine.

4. Transfer the contents to a stick blender cup. Add the salt, sugar and MSG. Purée the mixture. The sauce can now be stored for up to a week.

5. MAKE THE PASTA DISH. Defrost frozen porcini mushrooms (if you are using fresh porcini then ignore this step). Drain them on a sieve and reserve the liquid that you collect.

6. In a nonstick pan on a medium-high heat (#7 out of 10), add 2 teaspoons of bacon fat or vegetable oil, and a generous tablespoon of butter. When it foams up, add the drained mushrooms. Cook for 5 minutes.

7. Add the cooked mushrooms back to the liquid that was drained from them (if you are using frozen mushrooms - otherwise just set them aside).

8. Use a vegetable peeler on the celery. Then dice the celery and onion.

9. In a very large nonstick skillet, heat 2 tablespoons of butter on a medium heat (#6 out of 10). When the butter has melted add the onion. Spread it out on the pan, but do not stir the pieces.

10. In a separate pot, cook the pasta in salted water until al dente.

11. When there is some browning of the onions (6-8 minutes) stir and add the diced celery. Cook for about 3 minutes with occasional stirring.

12. Add the mushrooms and their juices to the pan. Cook for 3 minutes.

13. Now add about half of the Bear Sauce from above to the pan. Stir and continue cooking for another 3 minutes.

14. Now add the cooked pasta to the pan. Stir to warm through, then add the cream if you are choosing to use it.

15 When it tightens up (becomes dry), add some of the pasta water. Also some fresh herbs of your own choice. This dish reheats very well the next day, although of course that's not acceptable in Italian cuisine for pasta.

PIROZHKI (PIEROGI)
BAKED MEAT PIES

I titled this Pierogi in the video because that's what most foreigners call them, but the correct name is Pirozhki. An explanation of the naming follows after this recipe. Homemade pirozhki usually have a very simple filling. The amount of filling varies widely. Inexpensive commercial pirozhki often have meat thinned with bulgar wheat. These have a reputation for being made with cheap ingredients, so good restaurants usually steer clear of them. When they are offered at a top restaurant, the filling has to be exquisite, such as the example presented here.

FOR THE MEAT FILLING

600g (21 oz)	Beef Shank with marrow
100g (3.5 oz)	Carrots, diced medium
100g (3.5 oz)	Mushrooms, diced medium
60g (2 oz)	Shallots, diced medium
60ml (2 oz)	Brandy or Cognac
60g (2 oz)	Tomato Purée (pasata)
3/4 t	Coriander Seeds (or ground)
3/4 t	Black Peppercorns (or ground)
1	Bay Leaf, dried
small handful	Dill, fresh (no stems)
4 cloves	Garlic, peeled
2-3 T	Fat from charred steak (optional)

THE YEAST

These days the most commonly sold type of yeast is freeze dried. Very few professional bakers use this, though. Different strains of yeast impart slightly different flavors, and the strain that is bred to survive the freeze drying process best is not the strain with the best flavor. Use fresh!

PROCEDURE

1. MAKE THE MEAT FILLING. Begin by grinding the coriander seeds, bay leaf and black peppercorns to a powder in an electric spice mill. Stir the powder into the tomato purée.

2. Put some coarse salt on the beef. Heat a small stock pot on high (#8 out of 10). Add oil to the pan and brown the beef for 3-4 minutes on each side. Sear the sides of the meat some, as well.

3. Now remove the meat to a braising dish. Add more oil to the hot pan if it is running dry, and then add the carrots, mushrooms and shallots.

4. After about 3 minutes, add the tomato purée and spice mixture. Cook for 1-2 minutes with frequent stirring.

5. Now turn the heat off and add the cognac to deglaze the pan.

6. Add the dill and garlic cloves. Cook for about a minute.

7. Pour the contents of the pan over the browned beef in the braising dish. Put the lid on and braise in the oven at 160°C / 320°F for 4 hours.

8. When it is done, leave it to cool with the lid on it for an hour.

9. Pick the meat from the bone, being sure not to include any tough ligaments. Scrape the marrow out of the bone and put it in a food processor with the rest of the contents. Add 2-3 tablespoons of fatty charbroiled steak trimmings, if available. The amount you grind it is up to you. Russians want chunks left as proof of the quality of the ingredients that went into it. You can purée it if you prefer. Refrigerate overnight.

FOR THE DOUGH	
350g (8.8 oz)	Flour
75g (2.6 oz)	Butter
45g (1.5 oz)	Milk
30g (1 oz)	Yeast, fresh (see note on previous page)
1 t	Sugar
1 t	Salt
Additional flour for rolling out	
Egg Yolk (for glazing)	

11. MAKE THE DOUGH. Put the flour, butter and salt in the bowl of a

food processor. Run it in short bursts to combine, leaving small visible pieces of butter in the mixture (a coarse corn meal consistency).

12. Cover the mixture with cling film and put it in the freezer for 45 minutes to 1 1/2 hours (no longer).

13. During this time, make the yeast solution. Pour 200ml (7 oz) of hot water into a mixing bowl. Dissolve the sugar in it while it is still hot.

14. Add the cold milk to it to bring the temperature down. Feel it to make sure it is just barely above body temperature. If it is still too hot, then swirl it around some.

15. Add the yeast to the bowl, pressing against the sides to dissolve it. Cover the bowl with a damp cloth and wait 15 minutes.

16. Take the flour and butter mixture out of the freezer and put it in a large mixing bowl. Swirl the yeast mixture around (it should have foamed up by now) and add 150 grams (5.3 oz) of this to the flour. Discard the rest of the yeast solution. Stir to combine. Knead it some to make a dough.

17. Cover with a damp towel and let it stand at room temperature an hour.

18. Move it to the refrigerator (still covered with the towel) for 2-3 hours.

19. Divide the dough into two parts to make it easier. Roll it out on a

Dumpling Press

floured surface. Use a "dumpling press" (also called a gyoza press) to cut out circles and then encase the filling, as shown in the video. These are made of plastic and available inexpensively in specialty kitchen stores and online. The back side is a cutter and the top lets you seal in the filling.

20. Put a small amount of filling in each one. Fold to seal and place them on a silicone mat set on a baking sheet. Preheat oven to 190°C / 375°F.

21. Combine an egg yolk with 2 teaspoons of water and whisk. Brush each of the dumplings gently with the egg wash.

22. Bake them for 20 minutes first.

23. Now move each piece to a new position on the silicone mat (don't turn them over, though). Return tray to the oven for another 4-5 minutes.

24. Allow them to cool for at least 30 minutes on a wire rack before eating them. These go especially well as an accompaniment to Borscht (see recipe in Volume 1, page 44-45).

CONFUSION IN THE NAMING

"Pierogi" is one of the most confusing terms in all of gastronomy because there are several similar words that mean different, but related things. In Russian, a *pirog* is a large pie that can be savory or sweet. The plural of pirog is *pirogi*, often misspelled as pierogi—which is a Polish word for the boiled ravioli-like dumplings that Russians call *pelmeni*. When Russian pirogi are smaller in size, they are called *pirozhki*. Both the small and large type are traditionally made with a yeast dough and baked. Every Russian grocery store sells ready-to-eat baked pirozhki, as well as many different types of frozen pelmeni, which are taken home and boiled. Subtypes of pelmeni include *Siberian pelmeni* (which are little purses with a dough topknot that are traditionally filled with horse meat), and *varenyki* (which are also larger than pelmeni and filled with potato, fruit or a type of cheese that's similar to quark). All pelmeni may be pan fried after boiling.

POTATO PANCAKES WITH MUSHROOMS

Every Russian restaurant, including pizza and sushi places, has a menu that includes certain items for those who only eat Russian food. But it isn't Russian home cooking. It is in a category all by itself. This is not the type of potato pancake you are probably familiar with. They rich and soft on the inside.

300g (10.6 oz)	Potatoes, peeled
100g (3.5 oz)	Mushrooms, sliced
60g (2 oz)	Flour
45g (1.5 oz)	Butter
15g (1/2 oz)	Onion, sliced - ideally red onion
1-2 T	Dried Porcini Mushrooms (see note below)
1 teaspoon	Coarse Salt
1 whole	Egg
30ml (1 oz)	Smetana (or Crème Fraîche)
1/2 teaspoon	White Pepper, ground
2 T	Dill, freshly minced

Vegetable Oil, additional Smetana (or Crème Fraîche) for serving
Serve with pan-fried mushrooms (not part of this recipe)

DRIED PORCINI MUSHROOMS

You can use any dried wild mushrooms, but Porcini are best. You can either grind some first, or visually estimate the amount that will grind up into 1 to 2 tablespoons of powder in an electric spice mill.

PROCEDURE

1, Cut potatoes into large pieces and boil in lightly salted water for 8-9 minutes until tender but <u>not</u> mushy. Run cold water over the potatoes in a colander for a minute before leaving them to drain for about an hour. Note that the weight of potatoes called for in the recipe is after they have been cooked.

2. Heat a skillet on medium-high (#7 out of 10) and melt the butter in the pan. Now add the sliced mushrooms to the pan and sauté until they are turning golden. This should take about 5 minutes.

4. Grind the dried mushrooms and the coarse salt in an electric spice mill.

5. When the mushrooms have just started to turn golden, reduce the heat to medium (#5 out of 10) and add the sliced onion as well as the ground dried mushrooms. Stir. There will be a grainy feel for the first couple of minutes, then the mushrooms will exude moisture from the salt and the mixture will become softer. Cook 5-6 minutes in all. Add a few twists of fresh ground black pepper near the end of this time.

6. Transfer to a bowl and set aside until the potatoes are cooled.

7. In a food processor, combine the mushrooms, egg, white pepper, metana (or Crème Fraîche) and flour. Grind to a purée stage, scraping down the sides as necessary.

8. Now add the potatoes to the food processor. Pulse several times to break them up, then let it run for just a few seconds to make a homogeneous but very chunky mixture.

9. Transfer to a bowl and let the mixture to stand at room temperature for at least 15 minutes. You can keep it 2 days, but the cooking time is longer.

10. Heat a nonstick skillet on medium (#5 out of 10). Add enough vegetable oil to the pan to cover the bottom in a thin layer. You want to maintain the temperature around 143°C (290°F) during the cooking.

11. Put down a 10cm (4 inch) ring mold onto the pan and spoon out roughly 50 grams (3.5 oz) of the mixture into the mold. Smooth the top. As soon as it sets (less than a minute), remove the mold and repeat. You should get three of these in each pan. Cook for 3-4 minutes.

12. Carefully flip each one over and cook another 3-4 minutes.

13. Flip again. Cook for another 3-4 minutes.

14. Flip one last time. Salt lightly. Sprinkle with freshly minced dill. Cook about 3 minutes.

15. Serve with pan fried mushrooms and additional smetana on the side.

ATOMIC DUMPLINGS

Restaurants and nightclubs in Russia have a section of the menu devoted to "beer snacks". In less sophisticated establishment, these are usually very simple things like rye bread toasted with garlic oil. In better places, original and creative items are on the menu, such as this one that I developed for a club in Saint Petersburg.

300g (10.6 oz)	Potatoes, cooked (see notes below)
150g (5.3 oz)	Flour
25g (0.9 oz)	Buckwheat Flour
50-60g (2 oz)	Sardines, canned
60g (2 oz)	Peas, canned
45g (1.5 oz)	Onion, coarsely chopped
1 whole	Egg
2 teaspoons	Novgorord Seasoning (page 237)
1/2 teaspoon	Black Pepper, finely ground
	Cherry Poppins Tomato Sauce (page 124)
	Louisiana Crumble (page 245)

THE POTATOES

Ideally use Yukon Gold potatoes. Boil them in salted water until they are tender, but not mushy. Then leave then to cool on a colander for at least two hours until they are completely dry and room temperature before proceeding. This can be done the day before, in which case refrigerate the potatoes after they have cooled.

PROCEDURE

1. If you want to diminish the taste of the sardines, then soak them in water first. To reduce the taste even further, change the water two or three times over a period of half an hour. Either way, remove the spine bones

and pick out 50-60 grams of the cleaned fish.

2. In a food processor, combine the potatoes, flour, buckwheat, sardines, peas, onion, egg, black pepper and Novgorod Seasoning, if you are using it. Blend until it comes together as a slightly sticky dough.

3. Transfer to a bowl. Cover with a damp towel and let it rest at room temperature for 15-30 minutes.

4. Fill silicone molds with it, such as small ice cube trays (as shown in the video). Put these in the freezer for at least 3 hours.

5. In a small stock pot, heat 2 liters (2.1 quarts) of water with a tablespoon of salt until boiling.

6. Reduce the heat slightly and add the frozen dumplings. When the water returns to a boil, begin counting the time. Cook at a rapid simmer for 10 minutes, stirring occasionally.

7. Prepare au gratin dishes by spreading a layer of Charry Poppins tomato sauce (or plain tomato sauce) on each dish—enough to cover the bottom.

8. When the dumplings are done cooking in the water, remove them to a plate using a spider or slotted spoon.

9. After they have cooled briefly, arrange them on the au gratin dishes, flat-side down.

10. Broil at a distance of about 15cm (6 inches) for about 7 minutes until they start to turn golden on top.

11. Sprinkle on Louisiana Crumble generously. Return to the broiler for another 60-90 seconds. Allow to cool for a few minutes before serving.

FREEZER STORAGE

You can push the frozen dumpings out of the silicone mold into a plastic storage container and keep them in the freezer for at least two months. Don't try to store them after they have been boiled, though. Any time you want them, they are ready to go straight to step #5 above.

BEET PASTA WITH SMETANA

This is an example of a perfect natural fusion between two cuisines. Italians do make beet pasta sometimes (look it up if you don't believe me) but smetana and dill are not going to be the condiment. Conversely, beets are a traditional staple food for Russians and smetana is a normal condiment, but making pasta is quite rare. This is actually part of a more complex dish in which the pasta sits in a fish broth, The rest of this dish will be presented in a future volume.

90g (3 oz)	Beets, peeled and cooked
2 whole	Eggs
1 tablespoon	Olive Oil, extra-virgin
1/2 teaspoon	Salt
250g (8.8 oz)	Flour, regular all-purpose white

Purple Food Coloring (optional)
Smetana (substitute Crème Fraîche if you need to)
Dill, freshly minced

FOOD COLOR

Without additional food color, the pasta will be pink in color when it os finished cooking. That's not necessarily a bad thing, especially if you are using this as some kind or Valentine's Day dish, but a deeper purple will bring Borscht to mind better. If you can't get purple food color, you can mix red and blue together.

PROCEDURE

1. Chop the beets coarsely. Using a stick blender, purée the beets, eggs, olive oil, salt and food coloring (if you are using it).

2. Pass through a sieve. Discard any solids that won't pass.

3. Combine this with the flour using a food processor to make a dough. If it is too thick, add a little water. If it is too thin, add a little more flour. Be sure that it really needs these changes, though. The amount specified is usually right, though.

4. Let the dough rest in the refrigerator for at least an hour. Overnight is okay.

5. Use a pasta machine to roll it out and cut into fettuccine ribbons. As soon as each little bundle of pasta is cut, plunge it into simmering salted water for about 2 minutes to set it. Then drain it and toss in a bowl with unrefined sunflower oil (see Volume 2, page 74), or olive oil.

6. When you are ready to serve it, heat a little butter in a nonstick skillet. Toss the pasta with the butter to freshen it up and coat it.

7. Arrange in deep bowls and top with smetana and freshly minced dill. One advantage of high fat (30% or more) smetana is that it won't melt when it is on the hot pasta. It will retain it's shape and consistency until the person eating mixes it in.

SMETANA

Smetana is best compared to Crème Fraîche, but it is certainly not identical. Crème Fraîche generally has a fat content of 28%, while Smetana can vary between 10 and 40% fat. American style sour cream is usually in the 14% fat range. More important, the method of production is quite different and partly because of this and partly because of the high fat content, Smetana will not curdle when cooked and it won't melt when placed on hot soups or pasta dishes.

DAIRY-FREE CUPCAKES

Many bakeries in Russia do not add preservatives or anti-fungal chemicals. The problem is that the baked goods can start to turn moldy within as little as 72 hours. This is a serious logistical problem for product that has a very low profit margin to begin with. One day I witnessed bags of pita bread being opened and dumped into a tub. This was being recycled into cakes, such as this one. The result is much better than you might expect. I've heard of this being done in other countries, but it is probably more common in Russia where the use of preservatives is rare among independent bakeries.

130g (4.6 oz)	Flour
100g (3.5 oz) + 1 t	Sugar
70g (2.5 oz)	Pita Bread (commercial product)
60g (2.1 oz)	Corn Oil (see notes below)
3 whole	Eggs, separated
1 teaspoon	Baking Powder (not baking soda)
1/2 teaspoon	Food Coloring (optional)

ADDITIONAL NOTES

The surprising addition of pita bread adds both body and a slight salty earthiness that contributes to the surprising depth of flavor in this otherwise simple recipe that is reminiscent of sponge cake.

This can be made more flavorful by substituting half of the corn oil for pistachio nut oil. If you are using pistachio nut oil, I suggest a little green food coloring for the psychological benefit. Otherwise you can use any color you want, especially for children. Macadamia nut oil is okay, but more subtle.

PROCEDURE

1. Tear the pita bread into chunks and put it in a food processor. Grind it into crumbs. Now add the sugar and grind as fine as possible.

2. Preheat oven to 175°C (350°F). Add the flour, egg yolks, corn oil and the food coloring (if your are using it). Process into a homogenous mixture, scraping the sides and bottom of the food processor bowl as necessary. Don't mix this any more than you have to, though.

3. Add 120ml (4.3 oz) water. Blend to a paste consistency, but don't over blend it. Allow this to stand while you proceed with the other steps.

4. Using an electric mixer, whip the egg whites and the teaspoon of sugar until they reach firm peaks.

5. Gently fold in about a third of the flour mixture into the bowl with the whipped egg whites using a rubber spatula as if you were making a souffle. Now fold in another third, and then finally the last third.

6. Lubricate a silicone cupcake mold that has 6 large cups with a little corn oil in each cup.

7. Ladle in the batter to about 3/4 of the way up in each. Now bang the silicone mold gently against a table to make sure the batter is settled evenly.

8. Place the silicone mold onto a cookie sheet. Bake for about 28 minutes. At this point the top should just barely be starting to turn golden and the interior is fully cooked. Allow to cool for about ten minutes before turning them out.

9. Frost, if desired. Alternatively just sprinkle with powdered sugar.

> *"My diet plan is to make all of my friends cupcakes every week. The fatter they get, the thinner I look."*
>
> ...Yuliya Snigir (Russian actress)

"IT'S ONLY FOOD"

Kartoshka is one of the most popular fast food chains in Russia. The name means potato, and they specialize in potato dishes, as you might expect. When I ordered tater tots, they were prepared by microwaving the premade frozen product directly out of the freezer. The result was so bad that it was genuinesly funny. Slimy, sticky, goopy potato that was still cold in the center. When I took it back to the counter to complain that they should have deep fried them, they assured me that they were supposed to be microwaved. I couldn't believe that anyone would accept this as a food item, but the manager came out and swore that Russians ate this all the time and didn't care. Their excuse? Это только пища, or "It's only food." This is the alien mentality that's often exhibited toward food here and difficult to fathom until you have lived here for a long time.

Cooking in Russia

One of the questions that I am frequently asked is why am I living in Russia? This is especially baffling for Americans who have a distorted picture of what life here is like due to the media and the political agenda of past generations. Let me begin by dispelling a couple of common myths. First, the weather is about the same as central Ohio (where I also lived for a few years). Due to global climate change, there hasn't been any heavy snow now in five years. Besides, after spending years in boiling hot kitchens and 25 years in Los Angeles, I'm very glad to be in a cooler climate.

The next myth is that there are food shortages. I assure you that we have supermarkets that are just as large as anywhere else in the world and the shelves are fully stocked. I even put up a video showing what a Russian supermarket looks like.

Just as I am glad to be in a cooler climate, I'm also glad to not spend hours every week stuck in traffic. Public transportation here is *faster* than driving! It is completely safe and very inexpensive. Plus, there is no concern about drunk driving—either as the victim or the violator. When I go out to dinner, I don't want to be afraid that having one beer could turn me into a convicted felon.

One of the least obvious reasons that Russia was appealing was to step up to a tougher audience. After a few years in that hotel (see my book, *40 Years in One Night* for more about this), the compliments were perpetual jabber, and I was losing perspective. Without objectivity, you quickly turn into a self-indulgent fool who thinks everything he does is great. Especially in Hollywood where fake praise is nauseatingly normal, but also because these people loved food. They could see some good in any dish. The cure for that was a place where people *don't* love food, and any compliment is an illustrious achievement that's only uttered when it's heartfelt.

DELUXE ITALIAN TOMATO PASTE

This is a variation on the Real Italian Tomato Paste recipe presented in the YouTube video. This will produce even deeper flavor, but it requires a little more time and attention. The yield for this version is a bit more, too.

500g (17.6 oz)	Pasata, ideally San Marzano tomato purée
250g (8.8 oz)	Tomatoes, fresh (see note below)
25ml (0.8 oz)	Olive Oil, extra-virgin
1 teaspoon	Sugar
3/4 teaspoon	Salt

THE TOMATOES

Select the best quality tomatoes you can get. Preferably in season and vine ripened—and not too large or small. Out of season, cherry tomatoes can be used as a last resort.

PROCEDURE

1. Make a conical incision around the stem of each tomato and discard that part. The stem will not break down enough on cooking.

2. Chop the fresh tomatoes coarsely. Don't use a food processor! Pass the chopped tomatoes through the medium disc of a food mill. There is no substitute for this tool—if you don't have a food mill, you can't do it. It will take some effort to get it started. In the end you should have 40-50 grams (1 1/2 ounces) of pomace. Reserve the liquid and scrape out the skin and seeds from the bowl of the food mill to use in the next step.

3. Heat a heavy 2-liter stainless steel pan on a medium-high setting (#7 out of 10). Add the olive oil to the pan when it is hot. Wait 30 seconds, then add the pomace (skin and seeds from the bowl of the food mill). Stir

occasionally to keep it from burning, but not any more than you have to. Remember that constant stirring will inhibit caramelization. Within about 5 minutes you should see a good deal of fond sticking to the bottom of the pan. Keep cooking it as it darkens to a solid brown, but don't leet it burn.

4. Now deglaze with about a third of the passata. Also add the sugar and salt. Reduce heat to medium (#5 out of 10) and cook down slowly until the mixture is thick and dark. As in the video, allow a full minute to pass each time before you stir again.

5. Add the rest of the passata and the tomato liquid that passed through the food mill. Bring to a simmer, then reduce heat again to low (#4 out of 10). Stir about every ten minutes until the mixture is extremely thick. This may take two hours or more. Be patient. It needs to be a very thick bright red paste without any bitterness from burning.

6. Cool to room temperature and then store in a glass jar in the refrigerator. You should have about 250 grams (9 ounces) of tomato paste if you did everything right. Wait 2 days to use it for optimum results. It will keep for 1 to 2 weeks in the refrigerator.

THE RESULT

If you did everything right and you waited two days before sampling it, the taste will be intense, sweet, caramelized goodness and the flavor will a very long finish that will last on your tongue for several minutes after even a teaspoon size taste sample.

> *"The trouble with eating Italian food is that five or six days later you're hungry again."*
> ...George Miller

CHERRY POPPINS TOMATO SAUCE

You will need a food mill to make this. Do not purée it or you will destroy the balance. The name comes from the cherry tomatoes that pop like popcorn when cooked under pressure. The result is a sweet sauce with background notes reminiscent of fresh fennel. Consequently, this is an ideal tomato sauce to use with any fish recipe because the sweetness will play against the lemon that will likely be squeezed over the fish, and fennel and fish go together like apples and cinnamon.

500g (17.6 oz)	Cherry Tomatoes (see note below)
60ml (2 oz)	Olive Oil, extra-virgin
60ml (2 oz)	Water
30ml (1 oz)	Dry Vermouth
15ml (1 T)	Red Wine Vinegar
2 whole	Bay Leaves, dried
3/4 teaspoon	Salt
3-4 sprigs	Parsley, fresh

THE TOMATOES

Ideally use grape tomatoes and be sure that they are naturally ripe. Grape tomatoes, as they are often called, are not as sweet and the skin is thinner so they will pop easier, which is what you want here. Obviously this recipe can only be made when such tomatoes are available, but you can freeze the sauce for at least half a year with only a little loss in flavor.

PROCEDURE

1. Wash the tomatoes and put them into a pressure cooker.

2. Crumble the bay leaves on top. Add all of the rest of the ingredients and heat on medium-high (#7 out of 10) until it begins to boil.

3. Put the lid on and wait until steam begins to escape from the vent. Do not lower the heat. You want it at full pressure and venting during the entire cooking time.

4. When steam begins to escape, set a timer for 15 minutes. During this period you should hear tomatoes popping. If half the time has elapsed and there is hardly any popping, then increase the heat some.

5. When the time is up, turn off the heat. Leave the pressure cooker where it is at on the stove. Don't open it for another 30 minutes.

6. Now remove the parsley sprigs and discard. Pass the rest of the contents through the medium mesh plate of a food mill. Discard the solids.

7. Store in the refrigerator until needed.

MORE ABOUT THE CHEMISTRY

When the tomatoes pop they spray out the most volatile compounds first. These were the lowest-boiling point substances that built up the pressure to pop the tomato, after all. Within the pressure cooker at that moment is alcohol from the vermouth and olive oil. Many of these flavor molecules are not very soluble in water, but they are in alcohol or oil. That's the reason vodka is often added to tomato sauce—to help those compounds dissolve before they are lost in the steam. There is also a significant amount of bay in this, and as explained earlier in this book, bay releases isoprene which is a reactive substance that forms many different flavor molecules that we recognize as savory, herbal and spicy. The natural oil in parsley is rich in 1,3,8-p-Menthatriene, which has some structural similarity to the secondary flavor compounds found in fennel (which is a famous pairing with fish). These reactions are accelerated under acidic conditions, and so the addition of vinegar ensures that the pH remains low. As expected, there are overtones of fennel, but not the primary anise aroma—which is perfect for adding those fennel notes in a mysterious way. Of course you can use it with fennel, and double up on the chord.

ORIENTAL TOMATO CONCENTRATE

The term "oriental" is generally taken to mean Chinese and Japanese among Americans, but for Europeans this includes the Middle East and small republics like Kyrgyzstan. The tomato is a chameleon, able to have many different personalities and yet still be recognizeable as a tomato. Coaxing the right flavor profile out of tomatoes to go with a particular recipe is a key to making the ordinary into something extraordinary. This recipe gives you a whole new tomato flavor, and it's actually quite easy to make once you have the seasoning mix on hand.

150g (5.3 oz)	Tomato Purée (Italian Passata)
60g (2 oz)	Red Onion, finely diced
30ml (1 oz)	Olive Oil
3 teaspoons	Ottoman Spice Blend (page 238)
1 whole	Bay Leaf, dried
1/2 teaspoon	Sugar
1/2 teaspoon	MSG

HYPOCRISY?

Several times in this book I caution you not to use the "tertiary" spice blends at the back of this book as primary flavorings, so this may seem like I broke my own rule here. That's not the case though, because this tomato concentrate is not something you are going to eat by itself. It is itself an ingredient to be used in other recipes that (presumably) have stronger flavors going on in them already.

PROCEDURE

1. Heat a stainless or nonstick skillet on a medium heat (#6 out of 10). When it is hot, add the olive oil. Then add the red onion. Cook with

occasional stirring for 5-6 minutes.

2. Add 2 teaspoons of the Ottoman Seasoning and stir for about 30 seconds. Be ready with the next step, which will cool the mixture down before it starts to burn.

3. Add 120 grams (4 ounces) of the tomato purée and the sugar. Cook for about 8 minutes with occasional stirring until it is quite a bit thicker.

4. Add the rest of the tomato purée, the bay leaf and the rest of the Ottoman Seasoning. Lower the heat slightly (#5 out of 10) and continue cooking until quite thick. This should take about 5 minutes.

5. Add the MSG. Stir and cook for 1 more minute.

6. Remove the bay leaf. Transfer to a bowl to cool down rapidly at room temperature.

7. You can purée it, but this is seldom necessary for any application. The red onions should be quite soft and hardly noticeable as pieces by now. This should be stored in the refrigerator covered. It can be frozen, but it will lose some of its flavor after prolonged storage.

THE RESULT

If you did everything right, you should have just over 100 grams (3.5 ounces) of rich, dark tomato concentrate that's highly flavorful. Taste this and you'll immediately see how this can be used to elevate many recipes, especially Asian and Middle Eastern dishes.

UNORTHODOX APPLICATION

For a quick meal, try splitting a piece of pita bread, putting olive oil on the outside and this tomato concentrate on the cut surface to turn it into a type of pizza. Add some diced onion, mozzarella, grated parmesan and sliced chorizo sausage and bake at 200°C (390°F) for 7-8 minutes with fan assist ON. Reminiscent of the type of pizza frequently served in Croatia. Especially if it is baked in a wood burning oven.

SWEET & SOUR SAUCE

For a number of years, one of the finest Chinese restaurants in Los Angeles was Fung Lum's at the top of the hill at Universal City. Back in its glory days, the sweet and sour sauce was not the neon red goop that you usually find in Chinese restaurants. It was a magnificent deep sauce with layers of flavor. The secret was pork stock and partially thickening it by reduction instead of just loading it with corn starch. This is my copycat version of their sauce developed with some tips from one of their chefs.

FOR THE PORK BROTH

600g (21 oz)	Pork, cubed (see note below)
750ml (26.5 oz)	Water
1 T	Sesame Oil
20g (0.7 oz)	Garlic cloves, halved
1 teaspoon	Sichuan Peppercorns, whole
1 teaspoon	White Peppercorns, whole
small piece	Ginger, peeled
Coarse Salt	

THE PORK

The meat is used to make the stock. After that it is not used in this recipe, but it is perfectly flavored for making other Chinese dishes.

PROCEDURE

1. MAKE THE PORK BROTH. Wash and dry the pork. Season with coarse salt. Heat a pressure cooker on high (#8 out of 10). When it is hot, add the sesame oil and brown the pork well.

2. Make up a little bouquet garni bag with the garlic cloves, and both types of peppercorns. When the pork is browned, add the water, the lump

of ginger and the bag containing the garlic and peppercorns.

3. When it comes to a simmer, put the lid of the pressure cooker on. Adjust heat to keep it just below the venting point for 1 1/2 hours.

4. Take the pressure cooker off the heat and wait at least 30 minutes.

5. Open it and use a spider to remove the meat. Discard the ginger and bag containing the garlic and peppercorns. Refrigerate the broth overnight.

6. Skim the fat from the broth. It is now ready for use in the next part.

FOR THE SWEET & SOUR SAUCE	
1 recipe	Pork Broth (see above)
200g (7 oz)	Ketchup
75ml (2.6 oz)	Apple Cider Vinegar
75ml (2.6 oz)	Distilled White Vinegar
100g (3.5 oz)	Plums, fresh
1 whole	Orange, fresh - halved
1 whole	Lemon, fresh - halved
15g (0.5 oz)	Ginger, peeled and sliced
2 teaspoons	Corn Starch - dissolved in some cold water
2-3 teaspoons	Maggi Seasoning

7. MAKE THE SAUCE. Into a small stock pot, squeeze the juice from the lemon and orange. Add the fruit itself to the pot, too. Remove the pits from the plums and add them along with all of the other ingredients except the Maggi Seasoning. Bring to a simmer.

8. After 15 minutes, use tongs to squeeze out any liquid from the lemon and orange halves into the pot and discard the fruit. Maintain at a simmer.

9. After another 30 minutes, use a stick blender to purée the mixture.

10. After another 15 minutes of simmering, cool some and pass through a fine mesh sieve. Discard solids from the sieve.

11. Add 2-3 teaspoons of Maggi Seasoning to the sauce. Store in the refrigerator. You can freeze this for several months, if desired.

Footnote: Try mixing in 1-2 tablespoons of *Oriental Tomato Concentrate* (page 126) in at step 9.

MATHURINI BORDELAISE SAUCE

This is a versatile classic sauce for pork and game meats.

120g (4.2 oz)	Raisins, yellow
90g (3 oz)	Tomatoes, whole canned
30g (1 oz)	Shallot, coarsly chopped
250ml (1 cup)	Chicken Stock
2-3 T	Pork Demi-glace (see note below)
100ml (3.5 oz)	Red Wine, dry
15ml (1/2 oz)	Madeira
1 teaspoon	Cumin Seeds, whole
1 teaspoon	Juniper Berries
1/2 teaspoon	Coarse Salt
1/4 teaspoon	Black Pepper, finely ground
1 T	Butter
1 teaspoon	Beet Powder (see Volume 2, page 108)
1/2 teaspoon	MSG

PORK DEMI-GLACE

Although the best results will be obtained with the Pork Demi-glace recipe (page 150), you can substitute any gelatinous strong pork broth.

PROCEDURE

1. Warm the chicken stock up (either by microwave or in a pan on the stove). Dissolve the pork stock in with the chicken stock.

2. In a pan on a high heat (#8 out of 10) get a tablespoon of vegetable oil hot and then add the coarsely chopped shallot.

3. A minute later add the canned tomatoes. Cook for about 2 minutes.

4. The tomatoes should be mostly disintegrated now. Lower the heat to medium (#5 out of 10) and add the cumin seeds. Cook for 2 minutes.

5. There should be a lot of darkening and fond on the bottom, but nothing actually burnt. Add the raisins and continue stirring. Note that these are dried raisins that have not been soaked in any liquid beforehand.

6. Crush the juniper berries with the coarse salt in a mortar and add it to the mixture in the pan along with the pork and chicken stock mixture.

7. Bring it up to a simmer (still on medium heat) and scrape the bottom to pick up as much of the fond as possible.

8. Reduce heat to low (#2-3 out of 10) and maintain at a simmer 10 minutes.

9. Allow it to cool slightly before transferring to the cup of a stick blender. Add the dried beet powder and purée very well.

10. Pass through a sieve.

11. Rinse out the pan you have been using and add the sieved mixture back to the pan along with the red wine. Heat to a slow simmer.

12. When the sauce is reduced to the point that a spoon that's dragged across leaves a trail (see video) now add the Madeira and the black pepper. Continue heating until it is as thick as before you added the Madeira.

13. Remove the pan from the heat and whisk in the butter.

14. Stir in the MSG before tasting to adjust the salt.

MODIFICATIONS

As promised in the annotations at the end of the YouTube video, there are some optimizations that I suggest trying for specific game meats.

VENISON: Increase shallots to 60 grams (2 oz) and add 2 bay leaves at stage 6. Add 2 teaspoons of Texas Spice Blend (page 234) at stage 11.

WILD BOAR: Decrease chicken stock to 200ml (7 oz) and increase pork demi-glace to 90g (3 oz). Add a bay leaf at stage 6. At stage 11 add a full tablespoon of Burgundy Seasoning (page 230).

PHEASANT: Add 2 teaspoons of sherry vinegar and 3/4 teaspoon of allspice berries at stage 6. Use Sauternes in place of red wine.

GOOSE: Use dry white wine in place of red wine. Add 2-3 cloves garlic at stage 5. Add a teaspoon of dried sage at stage 11.

CHURRASCO AND CHIMICHURRI

Traditionally Churrasco is only seasoned with salt. The flavor comes from the cut and quality of beef along with being cooked over an open fire. This boosts the flavor and works on a vertical rotisserie, but cooking on a rotisserie over glowing coals is best.

500g (17.6 oz)	Steak - Angus Top Loin
1T	Brazilian Spice Blend (page 236)
1 teaspoon	Salt
CHIMICHURRI SAUCE	
30g (1 oz)	Green Serrano Chilies, seeded
30g (1 oz)	Cilantro, fresh
30ml (1 oz)	Lime Juice, fresh
20g (0.7 oz)	Vegetable Oil
2 cloves	Garlic
1 teaspoon	Brazilian Spice Blend (page 236)
1/4 teaspoon	Salt

CHIMICHURRI SAUCE

Most people agree that this is far better if it is made the day before, but no more than two days before. It's peak is about 24 hours after it's made.

PROCEDURE

1. MAKE THE CHIMICHURRI. Simply purée all of the ingredients in a stick blender cup (or regular blender if you are making a large amount). Cover and refrigerate until the next day.

2. Tenderize the steak using Jaccard devices, as explained in Volume 2 under Steak Mastery.

3. Cut the meat into large cubes so that as much of it as possible has a strip of fat on one face. Put it in a large bowl and rub in the Brazilian Spice and salt. Let it stand for 45 minutes to 2 hours at room temperature.

4. Thread the meat onto skewers. If you are using a vertical rotisserie ("barbecue maker") device, then put any pieces that do not have fat on them at the bottom of the skewer so that the fat from the pieces above them will drip down over them. This is one of the advantages of the vertical orientation design.

5. Cook them on a rotisserie until done. On the vertical rotisseries this will take about 14 minutes. Serve with the chimichurri sauce on the side.

MORE ABOUT THE BRAZILIAN SPICE MIX

Many of the seasonings in the *Spices & Coatings* section of this book (page 217) are intended as a subtle background note rather than being the sort of primary seasonings that you are used to. Trying to use these as the main flavor will generally not produce good results. A useful analogy for this is an advertisement back in the 1990's for Heineken beer in Germany. The slogan was that it makes you more of what you already are.

In one of these advertisements, a pirate with a peg leg and a parrot drinks the beer and now has two hook hands, two peg legs, two eyepatches and the parrot is now a menacing vulture. Use too much and that's what happens—but a little is a great enhancement.

HOLY MOLE!

This recipe comes with a stern warning. If you like spicy Mexican food, this is the equivalent of mainlining heroin! It is like nothing you have ever had before. It is way beyond merely delicious. Dice it up and put it on corn tortillas with Pico de Gallo for the best taco you've ever had, or even imagined.

1kg (2.2 lbs)	Pork Neck (also called Pork Throat)
160g (9 oz)	Onion, sliced - ideally a mix of white and red
100g (3.5 oz)	Tomato Purée (passata)
60ml (2 oz)	Orange Juice, fresh squeezed
30g (1 oz)	Garlic cloves, chopped coarsely
15g (0.5 oz)	Almonds, coarsely crushed
1 T	Corn Oil, or vegetable oil
1 T	Yucatan Spice Blend (page 223)
1 T	Corn Flour (fine corn meal)
1 T	Oaxacan Spice Blend (page 222)
1 T	Cocoa Powder (100% cacao)
50g (1.8 oz)	Chipotle Chilies, canned (see note below)
15ml (0.5 oz)	Red Wine Vinegar
15g (0.5 oz)	Dark Brown Sugar
20ml (0.7 oz)	Lime Juice, freshly squeezed
1 teaspoon	Coarse Salt

CHIPOTLE CHILI PEPPERS

These are actually smoked Jalapeño chilies (the name of some chili peppers is changed when they are dried or smoked). They are sold in cans and packed in a red Adobo sauce. If you prefer it very mild, then omit the chipotle chili, but you will be losing out on a lot of the flavor that way.

PROCEDURE

1. Slice the pork into four equal slabs. They should each be about 2.5cm (1 inch) slabs.

2. Combine together the Yucatan Seasoning, corn flour and coarse salt. Pour this over the slabs of pork and massage in. Allow to stand at room temperature for 15-45 minutes.

3. Set a <u>pressure cooker</u> on a medium-high heat (#7 1/2 out of 10). When it is hot add the corn oil. When the oil is hot, sauté the pork in two batches, cooking for about 3 minutes on each side of each piece (12 minutes total time spent, approximately). You want them very well browned, but not burnt. Remove the pork to a bowl to cool.

4. Add the sliced onion to the pan. Stir a little at first to get it spread out, then leave it alone for 2 1/2 to 3 minutes until it is almost burning.

5. Add the almonds and stir. Cook another minute.

6. Add the tomato purée, garlic and the Oaxacan Spice Blend. Stir more. Cook until it thickens, which should be about 2 minutes.

7. Add 200ml (7 oz) water and the red wine vinegar. Stir.

8. While this is reducing, cut the (now cool) pork into 2.5cm (1 inch) cubes. Add these to the pan and stir.

9. When the mixture is very thick again, add the orange juice. Reduce the heat to medium (#5 out of 10). Cook for about 3 minutes.

10. Add 100ml (3.5 oz) water. Then add the chipotle chilies in adobo sauce and the cocoa powder. Stir briefly. Put the lid on and reduce the heat to low (#3 out of 10). Don't worry about it coming up to pressure.

11. After 30 minutes, increase the temperature to medium (#5 out of 10).

12. After another 15 minutes, release the pressure and open it. Dissolve the dark brown sugar in 200ml (7 oz) of very hot water. Add the lime juice to this and pour it over the contents of the pressure cooker. Scrape that bottom to deglaze the fond. Add a dash of MSG if you are so inclined.

13. Now put the lid on and bring up to pressure for the last 15 minutes at a medium-high heat (#7 out of 10).

14. Cool it down, release the pressure and serve immediately, or store in the refrigerator for at least a week

SWEET AND SOUR PORK
OR CHICKEN, OR SHRIMP, OR COMBINATION

By far the most popular western Chinese restaurant dish,
though only vaguely related to actual Chinese sweet and sour.

220g (7.8 oz)	Pork, cubed (see notes below)
22g (0.7 oz)	Corn Starch
1 T + 1 teaspoon	Soy Sauce
1 T	Shaosing or Sherry
1 teaspoon	Sesame Oil
1/2	Egg (both yolk and white)
240g (8.5 oz)	Sweet and Sour Sauce (page 128)
90g (3 oz)	Red Bell Pepper, cubed
90g (3 oz)	Onion, coarsely chopped
90g (3 oz)	Pineapple chunks (canned)
45g (1.5 oz)	Green Bell Pepper, cubed
25ml (0.9 oz)	Peanut Oil
15g (0.5 oz)	Garlic, minced
15g (0.5 oz)	Ginger, minced or grated
2-3 T	Scallions, green part only
3/4 teaspoon	MSG

Vegetable Oil for frying, Sesame Seeds (optional garnish)

THE MEAT

Pork that has some fine grain uniform fat marbling like prime beef is ideal for this, but that can be difficult to find in the United States. In order for it to be tender, you need to aggressively tenderize it with a Jaccard device as explained in Volume 2, pages 182-185.

Chicken breast will also need to be tenderized in the same manner, though not as much. Chicken thigh and leg meat will work fine

Shrimp should be raw and between medium and large. The time to deep fry will be less and depend on the size, but around 90 seconds.

PROCEDURE

1. Cube the pork (or chicken) into bite size pieces about 2.5cm (1 inch) in size. Put the meat (or shrimp) in a bowl and mix with 1 tablespoon of the soy sauce, the Shaosing (or sherry) and the sesame oil. Mix well. Leave to marinate between 1 and 5 hours.

2. Whisk an egg. Add half of it to the pork (or other meat) along with the corn starch. Mix together well and let stand 15 minutes, mixing more occasionally. Heat the oil for deep frying to 170°C / 340°F.

3. Deep fry the pork (or other meat) a few pieces at a time so that the oil temperature doesn't drop much. Maintain it around 170°C / 340°F. Set the fried pork aside on paper towels to drain.

4. Put a very large nonstick skillet on a high heat (#8 out of 10). When the pan is hot, add the peanut oil to it and then reduce the heat just a little.

5. Fry the ginger and garlic for 1 minute.

6. Add the red and green bell peppers, the onion and the pineapple. Fry these with frequent (but not constant) stirring as long as you can before the garlic starts to burn. Around 5-6 minutes. This is the most difficult part—and no, you can't just add the garlic later. It has to go into the hot oil.

7. Add the Sweet and Sour Sauce to the pan and stir. Cook for 3 minutes.

8. Add the fried pork or chicken to the pan (shrimp goes in latter if that's what you are using). Stir to combine then put a lid on the pan. Reduce the heat to medium (#5 out of 10) and cook for 4-5 minutes.

9. Chop the scallions during this time.

10. Remove the lid and add half of the scallions. Also add the MSG and the last teaspoon of soy sauce. Stir to combine and keep cooking to reduce the sauce some and thicken it up. Taste and adjust the salt level. It will probably need a little more adding. When you judge that it is about a minute from being done, add the shrimp if you are using them.

11. Transfer to a serving platter and add the remaining scallions over the top. Consider adding a drizzle of sesame oil or some sesame seeds.

CHICKEN IN CHAMPAGNE PLUM SAUCE

This is a renowned Hong Kong hotel specialty dish.

PLUM SAUCE	
400g (14.1 oz)	Plums, pitted
350ml (12.3 oz)	Water
120g (4 oz)	Sugar
1 teaspoon	Cloves, whole
2 T	Ketchup
15g (1/2 oz)	Ginger, sliced
3/4 teaspoon	MSG, or 1/2 teaspoon salt

TABLESIDE PRESENTATION

This is cooked tableside on a chafing dish. Then the guests can see the champagne being added and then they are given the rest of the half-bottle.

PROCEDURE

1. MAKE THE PLUM SAUCE. Combine all of the ingredients except the MSG in a sauce pan. Bring to a simmer.

2. Cook for 1 to 1 1/2 hours until it thickens by reduction.

3. Strain through a fine mesh sieve. Stir in the MSG before bottling and storing in the refrigerator until at least the next day. The sauce will keep for at least a week in the refrigerator and a very long time in the freezer.

200g (7 oz)	Chicken Breast
15g (1/2 oz)	Corn Starch
30ml (1 oz)	Peanut Oil
200g (7 oz)	Plum Sauce
100ml (3.5 oz)	Champagne, Brut
1/2 teaspoon	Sesame Oil
to taste	Scallions, cut decoratively

4. MAKE THE CHICKEN. Cut the chicken into pieces on the bias. Put in

a bowl and coat with the corn starch. Leave to stand 15-30 minutes.

5. Heat a large skillet on a medium-high heat (#6 out of 10). When it is hot, add the peanut oil. Wait 30 seconds for the oil to get hot.

6. Add the chicken pieces to the pan. Cook with frequent stirring to get the pink gone from all surfaces.

7. Add the plum sauce to the pan. Note that the plum sauce recipe makes enough to do this recipe twice or more.

8. Stir and continue cooking to reduce the bubbling sauce until it is thick.

9. Add the champagne and continue stirring. Reduce until thickened some. It should be just a bit thicker than pancake syrup.

10. Transfer to a plate. Sprinkle with the sesame oil. Cut the scallions and garnish.

ALTERNATIVE VERSIONS FOR DUCK OR FROG LEGS

This is also well suited to duck, but the duck needs to be roasted first. It has also been used with deep fried frog legs, which are similar to chicken in flavor. If you are using either duck or frog, then add half the oil to the hot pan and jump to step 7. Add the duck or deep fried frog legs in at step 9 just to coat them in the sauce. Otherwise everything is the same.

PORK CUTLETS
WITH CAULIFLOWER CRUST

Cauliflower makes a good crust because it has a fairly neutral taste, doesn't absorb oil the way bread crumbs do, and has some good nutritional value. It works best with spicy foods, which compliment caramelized cauliflower in a novel way.

4 large	Breakfast Style Sausage Patties
	or 2 Pork Chops (see notes below)
120g (4.2 oz)	Cauliflower
40g (1.5 oz)	Shallots
3/4 teaspoon	Salt
15g (1/2 oz)	Butter
1 tablespoon	Flour
1 teaspoon	Beautiful World Spice Blend (page 233)
2-3 tablespoons	Louisiana Crumble (page 242)
Tabasco Sauce, fresh Cilantro or Basil	

THE MEAT

You can use boneless center-cut pork chops for this with good results, as I did in the YouTube video. Better results are obtained by using this with breakfast sausage pattie. If you opt to use plain pork chops, then use a Jaccard device on them first to make them tender as diagrammed in Volume 2 of this series under *Steak Mastery*.

PROCEDURE

1. If using pork chops, place a sheet of cling film over them and pound each steak gently until it is just a bit less than 10mm (3/8 inch) thick.
2. Dust both sides of the meat liberally with flour and allow it to stand while you prepare the coating.

3. You can use scraps of cauliflower from trimming florets, if you have enough, but don't use the tough stems. Put the cauliflower, shallots and salt into a food processor and grind until almost a paste. Scrape down the sides as needed.

4. Press the cauliflower mixture onto the top surface (one side only) of each pork escalope. Make sure it is pressed in well and flat. Let stand a few minutes and blot with paper towel before proceeding.

6. Heat a nonstick pan on a medium heat. When it is hot, add enough vegetable oil to lightly coat the pan. Swirl the pan to minimize the amount of oil you need.

7. When the oil is about 180°C (360°F), add one of the pork slices, cauliflower-side down. Cook them one at a time to ensure even cooking and so that the temperature of the pan does not drop too much.

8. After about 4 minutes, the coating side should be browned. Do NOT use a spatula to turn the meat, or you will destroy the coating. Use a fork gently at the edge of the meat to lift it up gently. The first time you do this you may need to adjust the heat (hotter if the coating is still white, or cooler if the coating is burning).

9. Cook on the reverse side 2 minutes.

10. Now flip it over for a final minute or so (again, use a fork and not a spatula). Remove to a plate and sprinkle with the dry herbs on the cauliflower side. The warm moisture will rehydrate them. Repeat the process with the other pork escalopes.

11. Stir the lemon juice into the Bechamel sauce with the nutmeg.

12. The meat is now fully cooked and may be set aside until it is needed. To finish, top with grated cheese and a dollop of the Bechamel. Put under the salamander (or broiler) to melt the cheese and brown the sauce slightly. Top with a minced fresh parsley or chives, as you prefer.

MARINATED BROCCOLI STEMS

If you love marinated artichokes, but not the price, then this recipe is going to become an instant favorite.

400g (14.1 oz)	Broccoli stems, trimmed and cut into chunks
40g (1.4 oz)	Garlic, chopped
40ml (1.4 oz)	Red Wine Vinegar
1 T	Salt
2 teaspoons	Red Pepper Flakes, preferably Italian
3-4 sprigs	Oregano, fresh
5-6 branches	Thyme, fresh
1 whole	Bay Leaf, Turkish (dried)
500ml (17.6 oz)	Olive Oil, extra-virgin
200ml (7 oz)	Vegetable Oil (or additional olive oil)
40ml (1.4 oz)	Lemon Juice, fresh
2 teaspoons	Lemon Zest, freshly grated

THE HERBS

You can choose other herbs such as marjoram, mint, basil and rosemary, if you prefer. The blend of oregano and thyme has been a consistent crowd pleaser, so make substitutions at your own risk.

PROCEDURE

1. Place the broccoli in a small sauce pan and add enough olive oil to cover completely. Now add the garlic, salt, red chili flakes, mint, marjoram and thyme.

2. Heat on a medium-low setting. Bring the temperature to 65°C (150°F).

3. Cover and reduce heat on the stove to low (#1 to 1 1/2 out of 10). Maintain between 75 and 85°C (185°F), checking with a thermometer occasionally and adjusting the stove as necessary.

4. After 1 hour add the lemon zest and lemon juice. Stir. Put the lid back on and remove from the heat. Allow the pot to stand undisturbed for

another hour, gradually cooling to room temperature.

5. Transfer to jars and store in the refrigerator for at least a week before opening to consume. Two weeks is even better, but one week minimum.

ADDITIONAL NOTES

The garlic cloves that were marinated in with the broccoli are also delicious and can be included on an antipasto plate. The oil (after you have used the broccoli stems) makes a great base for spicy salad dressing.

Broccoli came about from selective breeding, as did other members of the Brassica family that we now tend to think of as separate vegetables.

BRASSICA

FISH ROASTED WITH TOMATO SAUCE, FENNEL, CAPERS AND HOLLANDAISE

Baked fish with tomato sauce, capers and olives is a classic dish found throughout the Mediterranean. Hollandaise on the side is something you only find in France, though. The main difference here is the Cherry Poppins Tomato Sauce, which is the ideal flavor to go with this dish because it compliments the fennel and it has a natural sweetness as counterpoint to the lemon juice in the Hollandaise.

2 filets	Zander, Cod, or Walleye
1/2 to 3/4 teaspoon	Fennel Salt (see note below)
1/2 recipe	Cherry Poppins Tomato Sauce (page 124)
2 T	Capers, rinsed
2 T	Black Olives, sliced (optional)
60g (2 oz)	Fennel
FOR THE HOLLANDAISE	
60g (2 oz)	Butter
2	Egg Yolks
90ml (3 oz)	White Wine, dry
2 T	Shallots, minced
15ml (1/2 oz)	Apple Cider Vinegar
15-30ml (1/2 - 1 oz)	Lemon Juice, fresh
1/2-3/4 teaspoon	Fenugreek Leaves, dried or substitute dried Tarragon
30-60g (1 - 2 oz)	Clarified Butter (optional - see recipe)

CHOICE OF FISH AND SEASONING

You can use any firm, neutral tasting white fish for this. Another excellent choice is Chilean sea bass. Ideally you want fresh fish. Previously frozen fish has a lot of moisture in it. Use Fennel Salt (page 21) instead of table salt to season the fish. This was not in the video.

PROCEDURE

1. Trim the tough outer layer from the fennel bulb. Shave it thin (0.5mm) on a mandoline. Place this on a ceramic baking dish large enough for the fish.

2. Filet and salt the fish. Place the pieces on the shaved fennel. Spoon the Cherry Poppins tomato sauce over the top. Add the rinsed capers and some black olives (if you are using them).

3. Roast in a preheated 210°C (430°F) oven with fan assist ON for about 15 minutes.

4. MAKE THE HOLLANDAISE. While the fish is roasting, combine the shallots, white wine, cider vinegar and fenugreek leaves in a small sauce pan. Bring to a slow boil and reduce until it is getting close to dry.

5. Now add the butter to the sauce pan. After about a minute turn the heat off.

6. When the butter is nearly all melted, whisk it rapidly while adding the egg yolks. Whisk for about a minute to make sure it is emulsified.

7. Now add the lemon juice and a little salt to the pan and whisk some more. Leave it to stand, whisking only occasionally while the fish finishes cooking.

8. If you want a thinner, more traditional Hollandaise, you can thin it out with some clarified butter now that it has set up. Simply whisk the clarified butter in. The sauce won't break if you add it slowly. You can also pass it through a sieve now to strain off the shallots, if desired.

9. Transfer the fish to a plate. Either add the Hollandaise to the plate, or serve it on the side in a small cream pitcher. If you are serving it on the side, then definitely thin it out with clarified butter so that it will pour.

TORO GARDIANE
(BULL MEAT STEW)

This is ideal for restaurant service because (like many stews) it tastes best after being stored for a few days.

900g (2 lbs)	Bull Meat, ideally collar
500ml (2 cups)	Red Wine, dry
	ideally Costières de Nîmes from France
30ml (1 oz)	Red Wine Vinegar
1 whole	Red Onion, large (quartered)
2 large strips	Orange Peel (use a vegetable peeler)
2-3 branches	Thyme, fresh
2 whole	Bay Leaves
2-3 T	Ketchup
120g (4.2 oz)	Niçoise Olives, pitted and coarsely chopped
3-4 cloves	Garlic

Olive Oil and coarsely ground Black Pepper

BULL MEAT

If you can't get bull meat, you can make this with skirt steak. This was a peasant dish originally made in Spain from bulls killed in bullfights.

PROCEDURE

1. Trim and cube the bull meat. In a plastic container with a lid, combine the bull meat with the quartered red onion, the orange peel, the thyme, bay leaves, red wine vinegar and finally the red wine. Make sure there is enough wine to completely cover the meat. If not, add more. Put the lid on and refrigerate for at least 24 hours, and preferably for 3 days.

2. Now remove the onion and orange peel to a bowl to reserve. Discard the thyme and bay leaves

3. Strain the meat on a colander and collect the juices that run off. Let the

meat drain for 20-30 minutes before proceeding.

4. Put the juices that ran off into a blender with the red onion and orange peel. Add the garlic cloves to the blender and purée.

5. Pass the mixture through the medium-fine plate of a food mill. You will not get very good results with a sieve.

6. Transfer the liquid you extracted to a sauce pan and bring to a simmer on a medium heat (#5 out of 10). Watch that it doesn't foam up and overflow the pot.

7. Remove the meat from the colander and dry it well on paper towel. Season with a teaspoon of sea salt and a generous amount of freshly ground coarse black pepper.

8. Preheat oven to 200°C / 390°F. Heat a large skillet on a medium-high setting (#7 out of 10). Add about 30ml (1 oz) of extra-virgin olive oil to the pan. Swirl to coat the bottom evenly. Now brown the bull meat in the oil in two batches. Don't try to do it all at once or the meat won't brown properly. Use a splatter guard.

9. Transfer the meat to a braising dish. Deglaze the pan that the meat was browned on using a little of the blended red wine marinade you have simmering in the other pot. Add this to the braising dish, along with the rest of the simmering marinade.

10. Put the dish in the oven and then immediately lower the oven temperature to 165°C / 330°F. Keep the lid on it and let it cook for at least 3 hours. If it is not tender enough by that time, give it longer.

11. When the meat is tender, remove it to a bowl. Pour the liquid from the braising dish into a skillet and reduce it. When it starts to get thick, add the ketchup and the olives.

12. When it is quite thick, recombine it with the meat. Although you can eat it right away, it is traditional (and better) to refrigerate it and let it rest for at least a couple of days first. The flavor will be better and the meat will be more tender.

✦

147

ALIOTO MEATBALLS

From a San Francisco restaurant around 1960. For the complete history of this recipe, watch the video on YouTube.

600g (21 oz)	Steak (see note below)
100g (3.5 oz)	Pork, mostly fat
50g (1.8 oz)	Bread Crumbs
50g (1.8 oz)	Black Olives, pitted and drained
30g (1 oz)	Shallot
10g (1/3 oz)	Basil leaves, fresh (a small hand full)
1-2 cloves	Garlic, diced fine
1 whole	Egg
3/4 teaspoon	Salt
3/4 teaspoon	MSG

You will also need:
Flour, Olive Oil, Red Wine (ideally Amarone)

THE STEAK

The ideal beef for this recipe is charbroiled ribeye steak. If you eat steak fairly often, you can save scraps in your freezer until you have enough accumulated to make this. Alternatively, you will have to cook steak for the purpose of making this, as shown in the video.

PROCEDURE

1. Dry the meat off and season with coarse salt and coarsely ground black pepper. Either brown very well on a cast iron skillet, or (much better) cook over hot coals on a barbecue. You want deep color on the outside, but still pink inside—what is called black and blue.

2. Let the meat cool to about room temperature. Now wrap it up tightly in cling film and refrigerate until the next day.

3. Cut the steak into pieces that can go through the meat grinder. Same thing for the pork. Choose the fattier parts of the pork. Grind together two

imes on the medium fine perforated plate.

4. Dice the garlic small and set it aside for a few minutes, as explained in Volume 2, on page 51.

5. Using a food processor, combine the bread crumbs, olives, shallot, basil, garlic, salt and MSG. It helps to chop up the shallot and basil some first before adding it to the food processor.

6. When the mixture is fairly homogenous, add the egg and blend again.

7. Combine this mixture with the ground meat. Cut it together at first using a spatula, and then knead by hand.

8. Sprinkle flour down on a large plate or other work surface. Roll meatballs by hand (about 30 grams or 1 ounce each). Place the meatballs on the floured surface. Then sprinkle with more flour and roll them around a little to coat all of the sides with a little flour. You don't want them dredged in flour, though—just a light coating.

9. Add enough olive oil to a hot nonstick skillet to coat the bottom. The heat should be medium-high (#6 out of 10). Swirl the pan to coat evenly, then fry the meatballs in batches. At first have the meatballs in the middle of the pan where it is the hottest to sear the edges. Then move them into a ring around the edge of the pan where the heat is the most even. Turn them occasionally, keeping them along the outer edge, until they are browned.

10. Move the meatballs away from the edge to form a ring in the mid-center of the pan. Add half a glass or so of Amarone and put a lid on. Lower the heat some (#3 out of 10). Wait for 2-3 minutes.

11. Now remove the lid and turn the meatballs over in the wine sauce which was thickened by the flour that was on the outside of the meatballs). When they are evenly coated, put the lid back on and cook for another 3 minutes. They should have absorbed virtually all of the wine at this point. You can serve them immediately, or store them and reheat later, but they are at their best when consumed within a few hours of cooking.

✛

PORK DEMI-GLACE

Although pork stock is common in Asia and Mexico, it is almost nonexistent in European cuisine. There are many proposed explanations. Probably the most likely is that stocks are mostly used to make sauces—the food of nobility. Pig meat was mostly seen as peasant food. Examine the recipes of old European nobility and you'll see very few pork recipes. The exception is roast suckling pig, but that doesn't need a sauce.

2 kg (4.4 lbs)	Pork, ideally meaty pork ribs (see video)
2-3 T	Vinegar
2 T + 1 teaspoon	Ottoman Spice Blend (page 238)
150g (5.3 oz)	Tomato Purée (passata)
75ml (2.7 oz)	Red Wine, dry
30ml (1 oz)	Vegetable Oil
2-3 cloves	Garlic, crushed
1 1/2 teaspoons	Sugar
1 medium	Onion, coarsely chopped
1 medium	Carrot, coarsely chopped
1 large	Celery stalk, coarsely chopped

PROCEDURE

1. MARINATE THE PORK. In a plastic container with a lid large enough to hold all of the pork, add the vinegar and then enough water to cover the meat. The water should be just above room temperature. Let stand 1 hour.

2. Preheat oven to 220°C (430°F). Rinse the meat off with cold water and then dry it. Rub 2 tablespoons of the Ottoman Spice Blend into the fatty top side of the meat. Roast in a heavy bottom pan for 1 to 1 1/2 hours.

3. While the meat is roasting, make the tomato concentrate. Begin by heating a pan on a high heat (#9 out of 10). When it is very hot, add the vegetable oil. Wait about 30 seconds for the oil to get up to temperature,

then <u>carefully</u> add about half the tomato purée, shielding yourself with a splatter guard. Clamp the spatter guard down as soon as you can.

4. As soon as the mixture stops erupting violently and you can remove the splatter guard safely, scrape out the rest of the passata into the mixture and add the sugar, too. Do not stir ir. Put the splatter guard back and reduce the heat to medium-high (#7 out of 10). Cook for 2 minutes.

5. Now remove the splatter guard and stir the mixture. Add the garlic and stir for 10 seconds.

6. Turn the heat off. Add the red wine, scraping the bottom to deglaze the pan. Then add the other teaspoon of the Ottoman Spice Blend. Move the pressure cooker off the burner and let the residual heat finish the cooking.

7. When the pork is finished roasting, strain off the fat and save it for other applications. You don't need it for this. Now put the chopped carrot, onion and celery in the bottom of a pressure cooker. Break up the pork ribs and put those on top, then the tomato concentrate mixture you made. Add water up to the fill line on the pressure cooker. Bring to a simmer.

8. Now put the lid on and get it up to pressure. Now reduce the heat to keep it just below the point that it is venting steam for the next 1-2 hours.

9. Let the pressure cooker cool down without opening it for 30 minutes.

10. Remove the meat. You can use it for something else, but it may have too many bones and not enough flavor to be worth your time picking through it, depending on what cut you used. At any rate, pass the rest of the liquid through a sieve. Discard the vegetables. Refrigerate overnight.

11. Skim the layer of fat from the top. You can use the fat for something else, too. It isn't needed here. Now simmer the liquid to reduce it further. Do <u>not</u> boil it. Reduce it slowly at about 80-85°C (180°F).

12. When it has been reduced to about 200 grams (7 oz), pass it through a sieve. Store in the refrigerator for up to a week. It will keep for months in the freezer. Unlike ordinary demi-glace, this is more about the flavor than the gelatine thickening power.

RICH LAMB BROTH

This is not to be confused with lamb stock. Remember that broths are seasoned and the emphasis is on the meat, while stocks are neutral and the emphasis is on the bone gelatine (see page 94 for more about this).

1 kg (2.2 lbs)	Lamb shanks and/or ribs
1 whole	Dried Sweet Red Chili (see Volume 1) or 1 T Paprika
2 T	Dark Brown Sugar, ideally Cassonade
2 teaspoons	Black Peppercorns, whole
1 teaspoon	Coarse Salt
3 T	Wild Mushroom Paste (see notes below)
1 whole	Onion, medium
1 head	Garlic, whole
2-3 stems	Parsley, fresh
2-3 stems	Sage, fresh
2-3 branches	Rosemary, fresh
2-3 branches	Thyme, fresh

WILD MUSHROOM PASTE

The best choice for this is wild mushrooms that you have cooked down yourself, as explained on page 34. You can also purchase bottled porcini mushroom paste, but it is quite expensive. The next step down from there is adding some dried wild mushrooms. You don't need to bother reconstituting them with water, because that will happen automatically during the cooking process. Finally, as a last resort (which is actually still quite good, believe it or not) as shown in the video, you can substitute this ingredient with one whole Maggi brand mushroom stock cube. This is normal practice in a restaurant to save money because wild mushroom paste is very expensive and it is impractical to send cooks out foraging enough mushrooms during the brief season to last the rest of the year.

PROCEDURE

1. Preheat oven to 210°C (400°F). Crumble the red chili into an electric spice mill, but leave off the stem. Add the brown sugar, black peppercorns and coarse salt. Grind to a powder.

2. Rub the lamb with 2-3 tablespoons olive oil. Now pour the ground spice over the top and massage them in.

3. Roast in the preheated oven for an hour. Do not use a thin metal roasting pan for this, or the lamb will burn on the bottom. Use either a ceramic dish, a glass baking dish, or a heavy Dutch oven without a lid.

4. Slice the onion and the head of garlic and put it at the bottom of a pressure cooker. Place the roasted lamb meat on top of that, as well as all of the juices and fat from the roasting pan. Add the wild mushroom paste, or whatever you are substituting. Remember that this is a lamb broth and not a mushroom broth, so don't go overboard and add a lot. This is where the Maggi cube has an advantage, because one cube is the perfect amount. Now add 600ml (2 1/2 cups) water to the pressure cooker and all of the fresh herbs. You can substitute a teaspoon of dried herb for any that you don't have fresh, if you need to.

5. Close the cooker and bring it up to pressure. Maintain the heat so that it is just below the point where steam is venting for the next 1 1/2 hours.

6. Now move it off the heat, but do not open it for 2 more hours.

7. Remove the meat to a storage container. You can use this for soups or other applications. Discard the mushy herbs. Pass the liquid through a sieve and discard the solids from the sieve, too.

8. Refrigerate the collected broth overnight;

9. Pick off the solidified fat from the top. You can save it for cooking, as explained earlier (see page 91). The broth is now ready for use. It can be served as a soup just as it is. You can add other ingredients to make it into a proper soup (see next page), or use it as the base for sauces.

GREEK LAMB SOUP
WITH BEANS AND FETA

This is a simple soup to make once you have the Rich Lamb Broth recipe from the previous page made. The cooked lamb meat in this soup is a byproduct of that same recipe.

450g (16 oz)	Rich Lamb Broth (page 152)
120g (4.2 oz)	Lamb, cooked (see note below)
150g (5.3 oz)	White Beans, canned
75g (2.7 oz)	Rutabaga (see note below)
30ml (1 oz)	Olive Oil
150g (5.3 oz)	Tomato Purée (pasata)
1 teaspoon	Sugar
1-2 teaspoons	Ottoman Seasoning (page 238)
4 cloves	Garlic, cut in small pieces
Feta Cheese	
Parsley, fresh	

ADDITIONAL NOTES

Rutabaga is also known as yellow turnips, or Swedes. I suggest you use rutabaga for this, even if you think you don't like it. The overall flavor is different than eating a plate of rutabaga by itself. Still, you can substitute carrots if you feel you must, in which case simmer then only about 10 minutes instead of 15 because they are not as tough as rutabaga.

For the meat, you can use the lamb that was used to make the broth.

The Ottoman Seasoning was not mentioned in the video. You can omit it, but obviously it brings more layers of flavor to the dish.

PROCEDURE

1. Peel and dice the rutabaga neatly into 1.25cm (1/2 inch) cubes.

2. Simmer the diced rutabaga in lightly salted water for 15 minutes.

3. Cut the lamb meat into small cubes. Trim any gristle from the meat.

4. Heat a small stock pot on a high setting (#8 out of 10). When the pan is very hot, add the olive oil to coat the bottom (swirl to coat evenly).

5. Now carefully but quickly add about half of the tomato purée into the oil, using a splatter guard to shield yourself. Then cover the pan with the splatter guard.

6. As soon as the initial violent splattering is over, add the sugar. Put the splatter guard back and cook for another 30 seconds.

7. Now add the lamb. Spread it out on top of the liquid, stirring it as little as possible. Try not to disturb the bottom layer. Cook for another minute.

8. Now add the rest of the tomato purée and the Ottoman Seasoning (if you are using it) on top of the lamb. Reduce the heat slightly to medium-high (#7 out of 10). Put the splatter guard back and cook for another 3 minutes.

9. Now stir the mixture and add about 100ml (3.5 oz) of the lamb broth. Scrape the bottom to deglaze it. Bring to a boil.

10. Now reduce the heat to medium (#4 out of 10) and maintain at a simmer with occasional stirring for roughly 10 minutes until it is very thick and most of the moisture has evaporated.

11. Add the garlic and cook for 2 minutes with frequent stirring.

12. Add the rest of the lamb broth, the beans and the rutabaga. Bring to a simmer. Taste to adjust the salt. If it is too thick for your taste, then add water to thin it to your liking. Simmer for about 10 minutes in all at this point.

13. Spoon into bowls and crumble feta cheese on top. Add a little freshly minced parsley and a little fresh ground black pepper.

DHABA EGG CURRY

Dhaba refers to food sellers along the main roads of India. They offer home style dishes that are mostly vegetarian and are the staple diet of truck drivers. Egg Curry probably became so popular because it's often the closest thing to meat you can get.

4 whole	Eggs, hard boiled and peeled
120g (4.2 oz)	Tomatoes, fresh
120g (4.2 oz)	Tomato Purée (pasata)
100g (3.5 oz)	Red Onion, grated (see note below)
60ml (2 oz)	Ghee or Macadamia Nut Oil
2 teaspoons	Coriander Seeds
1/2 teaspoon	Cumin Seeds
6 whole	Green Cardamom pods
6 whole	Black Peppercorns
2 whole	Bay Leaves
5 cm (2 inch)	Cinnamon stick
15g (1/2 oz)	Garlic, chopped fine
15g (1/2 oz)	Ginger, peeled and grated
1-3 teaspoons	Kashmiri Mirch (hot red chili powder)
1 teaspoon	Methi (dried fenugreek leaves) or cilantro
1 T	Butter (optional)
1 teaspoon	Brown Sugar (optional)
1/2 teaspoon	Garam Masala (see Volume 1, page 134)
Fresh Cilantro and Shallot for garnish	

GRATED ONION

The weight here is after grating, not before. Also, shallots are traditional, but red onion is less expensive and will work okay because of the other strong flavors going on..

PROCEDURE

1. Purée the fresh tomatoes with the passata using a stick blender.
2. Heat a small saucepan on a medium-high stove setting (#7 out of 10).

When it is hot, add the oil or ghee and let it get hot.

3. Use a knife to poke a deep hole in each of the hardboiled eggs before adding it to the hot oil. Fry until they are golden brown on all sides. This will take about 8 minutes. Use a splatter guard if it starts spitting.

4. Add the coriander seeds, cumin seeds, cardamom pods, black peppercorns, bay leaves and cinnamon stick to the hot oil left in the pan. Cook for about 15-20 seconds while scraping the bottom of the pan to loosen the fond.

5. Now add the grated onion and reduce heat to medium (#4-5 out of 10).

6. After about 3 minutes add the ginger and garlic. Cook 3 minutes more.

7. Now add the tomato passata purée mixture and 3/4 teaspoon salt. Continue cooking with occasional stirring for another 5-6 minutes.

8. Add the Kashmiri Mirch. Continue cooking for another 20-25 minutes.

9. Now thin it with 200ml (7 oz) water. Stir to combine, then transfer to a blender. Purée <u>well</u>, and then pass through a sieve. Discard any solids.

10. Rinse out the pan you were using before and return the sieved curry sauce to the same pan. Bring back up to a simmer on a medium heat.

11. By now the eggs should be cool. Cut them in half and add them to the simmering curry. Add the methi over the top. Cover and cook 10 minutes.

12. Remove the eggs along with whatever curry sauce sticks to them to a platter. Add the butter and brown sugar to the remaining sauce in the pan. Heat and whisk together. Add the garam masala. Adjust salt to taste.

13. Pour the sauce over the eggs and garnish with cilantro and thinly sliced shallots.

VARIATION

For a more attractive presentation, you can continue making the sauce without putting the eggs back into it. Then spoon the curry sauce into a serving dish and place the eggs, cut-side up, into the sauce. They will not be as flavorful or soft this way, but the dish will look much prettier.

SALMON PAELLA
WITH CHORIZO AND GREEN OLIVES

Not traditional, but popular in Spain in recent years and with many variations, including even as tapas plate in Barcelona.

250g (8.8 oz)	Calasparra Rice (see note below)
750ml (26,5 oz)	Chicken Broth
200g (7 oz)	Salmon, large cubes
50g (1.75 oz)	Chorizo, Spanish
50g (1.75 oz)	Green Olives, chopped (see note below)
180g (6.3 oz)	Tomatoes
70g (2.5 oz)	Tomato Purée (pasata)
100g (3.5 oz)	Red Onion
5-6 cloves	Garlic
2 teaspoons	Pimentón (Spanish smoked paprika)
2 pinches	Saffron threads
Fresh Oregano	
Olive Oil	

THE RICE

You can substitute Bomba rice from Spain with very good results. If you make this with Arborio rice, as many sources claim is an acceptable substitute, you won't be able to get the deliciously crisp caramelized base to develop before it burns. It'll still be a tasty dish, but it won't have the characteristic crust that is unique to Paella.

THE OLIVES

I suggest using Schiacciate olives from Calabria, Italy. These are flavored with oregano, chilies and dried tomatoes. Gordal olives from Spain are also good. Pit them with an olive pitter or the side of a knife.

PROCEDURE

1. Slice the chorizo into bite sized pieces.

2. Heat a large skillet or paella pan on a medium setting (#4 out of 10). Add enough olive oil to coat the bottom of the pan. Swirl to coat. Add the chorizo and fry for about 4 minutes to flavor the oil and get it a little crisp.

3. While it is cooking, coarsely chop the tomatoes, red onion and garlic. Put those into a blender, or the cup of a stick blender, along with the tomato purée. Add about 2 tablespoons of fresh oregano leaves (or substitute a generous teaspoon of dried oregano). Purée.

4. Stir the smoked paprika in with the chorizo and oil. Cook 1 minute.

5. Now stir in the contents of the blender. Increase the heat to medium (#5 out of 10).

6. Meanwhile on another burner, heat the chicken broth up to a slow simmer. Put the oregano branch in with the broth with whatever leaves it has left on it from step 3. Crumble the saffron into the same pot.

7. When the sofrito has been cooking 15-20 minutes, it will be darkened some and not as liquid. Stir the rice into the sofrito. Cook for 5 minutes.

8. Add about 3 ladle fulls (170ml or 6 oz) of the chicken broth. Stir it in and even out the mixture so that it is flat. This is the last time you stir it!

9. After about 10 minutes the liquid should have been absorbed and it is dry on top. Begin adding liquid from the pan with the simmering chicken broth to it, a little at a time. Focus on the spots that look the driest. Continue doing this for. Do <u>not</u> stir it. Rotate the pan on the stove to move any puddle at the edge to the other side. Continue this another 10 minutes.

10. Now scatter the olives over the top. Then the salmon, cut into cubes.

11. After 2-3 more minutes add whatever remaining chicken broth you have left. Let it cook until it starts to become dry around the edges again.

12. Now put a lid on the pan and increase the heat to high (#8 out of 10). Cook for 4-5 minutes this way.

13. Remove the lid and lower the heat to low (#3 out of 10). Add freshly minced oregano. Continue cooking another 2-3 minutes.

14. Serve with wedges of lemon and garlic aioli.

ROTISSERIE GARLIC CHICKEN RICE
USING "BARBECUE MAKER" VERTICAL ROTISSERIE

A vertical rotisserie (also called a "Barbecue Maker" in many parts of Europe) is a great kitchen tool that can be used to produce evenly cooked kebabs better than you can in a broiler, although still not as good as over coals with a real rotisserie.

2 whole	Chicken Breasts, skinless
25g (0.9 oz)	Garlic cloves
1 T	Coarse Salt
1/2 teaspoon	Black Peppercorns
1-2 T	Olive Oil
FOR THE RICE	
500ml (18 oz)	Chicken Broth
240g (8.5 oz)	Basmati Rice, rinsed
100ml (3.5 oz)	Light Cream
1-2 whole	Red Chilies
4 T	Basil leaves, chopped
2 T	Parsley leaves, chopped
1 T	Campari Liqueur (optional)

PROCEDURE

1. Coarsely chop the garlic and put it into a mortar with the coarse salt and the black peppercorns. Crush this, making sure that all of the black peppercorns are broken up. Then add the olive oil and grind it some more.

2. Cut the chicken into large pieces (suitable for threading on the skewers), taking care to remove tendons. Put these in a bowl and add the garlic paste. Massage the paste into the chicken. Cover with cling film and refrigerate for 2-3 hours. You can leave it a bit longer, but if you leave it overnight it won't be as good. Try to time it for 2-3 hours from cooking.

3. Thread the chicken onto the skewers of the rotisserie.

160

4. Either cook the skewers on the vertical rotisserie that I mentioned at the start, or (even better) over hot coals using a horizontal rotisserie. On the "barbecue maker" they will take 14-16 minutes. You can just eat it now, if desired. The rest of this recipe is just one example of how to use it.

5. MAKE THE RICE. Cut the chicken into smaller pieces, if desired. Chop the red chilies, removing the seeds and membranes if you want it less spicy. Combine the chicken, rinsed rice, cream, chilies, basil, parsley, chicken stock and Campari in a braising dish. Cover and cook in the oven at 160°C / 320°F for 3 hours. Do not stir it during the cooking time.

6. Transfer it to a large bowl and fluff up the rice some with a fork.

FOR THE GARLIC AIOLI

1 whole	Egg Yolk
3-4 cloves	Garlic, crushed
1 T	Lemon Juice, fresh
Olive Oil, Salt	

---

7. MAKE THE AIOLI. Use a garlic press to crush the cloves into a bowl.

8. Wash the outside shell of the egg with warm soapy water. Read in Volume 2, page 218 to know why. Now separate the egg and put the yolk in the bowl with the garlic. You don't need the white for this recipe.

9. Begin whisking, adding a little olive oil at a time to form the emulsion. See the video if you are not sure how to do this. Add as much as you like to make the desired thickness. You can also add some commercial mayonnaise to this, too. Not only to save money (which is the main incentive to restaurants) but also because commercial mayonnaise contains stabilizers and some flavorings you can't buy.

10. When it is the consistency you like, whisk in the lemon juice. Adjust the seasoning with salt and possibly a pinch of sugar.

11. Plate it up with the garlic aioli, chicken rice, fresh basil and chilies.

THAI DUCK AND SPINACH SALAD

Based on a menu item at Saladang in Pasadena, California.

FOR THE DRESSING

45ml (1.5 oz)	Duck Stock (or chicken)
30ml (1 oz)	Orange Juice, fresh
30ml (1 oz)	Thai Fish Sauce (Nam Pla)
30g (1 oz)	Dark Brown Sugar (Cassonade)
22g (3/4 oz)	Garlic, chopped
22g (3/4 oz)	Ginger, grated
15ml (1/2 oz)	Soy Sauce
15ml (1/2 oz)	Oyster Sauce
1/2 whole	Star Anise
1 teaspoon	Corn Starch
22ml (3/4 oz)	Sesame Oil
22ml (3/4 oz)	Lime Juice, fresh
2 teaspoons	Sriracha Chili sauce

PROCEDURE

1. MAKE THE DRESSING 1-3 DAYS AHEAD. Mix the corn starch with the orange juice to make a slurry. Crush the star anise in a mortar. Put both of these into a small sauce pan along with the brown sugar, garlic, ginger, soy sauce and Thai fish sauce. Simmer until it thickens some.

2. Add the oyster sauce and continue simmering for about 10 minutes.

3. Transfer the contents of the pan to a glass jar with a lid. Add the lime juice, sesame oil and Sriracha. Put the lid on. Shake to combine and then refrigerate for 1 to 3 days. You can make a larger batch of this and store it in the freezer for weeks, if desired.

4. MAKE THE DUCK SEASONING. In an electric spice mill, grind together the cloves, cinnamon, fennel seeds, Sichuan peppercorns, white peppercorns, dried chilies, star anise and coarse salt. Adjust the amount of chilies to your own personal preference for heat.

FOR THE DUCK

2 large	Duck Breasts, skin-on
1/2 teaspoon	Cloves (the spice)
1/2 teaspoon	Cinnamon, ground
1/2 teaspoon	Fennel Seeds
1/2 teaspoon	Sichuan Peppercorns
1/2 teaspoon	White Peppercorns
1-4 whole	Dried Piri-Piri or Thai Chilies
1 whole	Star Anise
3/4 teaspoon	Coarse Salt

FOR THE SALAD

450g (16 oz)	Baby Spinach, fresh
6-8 large	Mushrooms, sliced 3mm (1/10 inch) thick
2 T	Lime Juice, fresh
16 whole	Cherry Tomatoes
6-8 cloves	Garlic, sliced thin

Sesame Seeds and Bean Sprouts (optional)

5. Score the skin of the duck breasts using a sharp knife. Put the duck in a bowl and massage the spice in. The them rest at room temperature 1 hour.

6. Either work in batches, or use multiple nonstick pans. Put the duck breast skin-side down in a cold pan. Begin heating it on medium (#6 out of 10). Put a splatter guard down if you have one. Cook for about 9 minutes.

7. Flip it over and cook for 2 more minutes. Preheat broiler

8. Drain the fat to a cup for later use. Broil the duck for 4 minutes, skin-side up about 15cm (6 inches) from the heat source. Then let it cool.

9. Slice the mushrooms and squeeze the lime juice over them..

10. Now purée the dressing and pass it through a sieve. Add the juices that ran off of the cooked duck breasts to the dressing. This is all optional.

11. In a large nonstick skillet, heat a tablespoon of the reserved duck fat per salad on a medium heat (#6 out of 10). Cook the garlic in the fat.

12. Add the cherry tomatoes and cook until soft. Then put in a bowl.

13. Add a little more duck fat and sauté the spinach until slightly wilted.

14. Warm the dressing in the same pan. Assemble the salad. Add dressing.

EYE OF ROUND STEAK
WITH IMPROVISED SOUS VIDE METHOD

Eye of Round is an inexpensive cut of meat, but it tends to be quite tough. Here is a method to make it tender that also demonstrates how you can imitate sous vide cooking without any specialized equipment. However, you will need both the single row and triple row Jaccard devices to do this correctly (see Volume 2, page 182). The video also shows this method.

Eye of Round Steaks, about 4cm / 1.5 inches thick
Garlic cloves, fresh (one per steak)
Rosemary, fresh
Thyme and/or Basil, fresh
Olive Oil
Butter
Fresh Cracked Black Pepper
Rich Beef Broth (see notes below)

RICH BEEF BROTH

You will need about 60ml (2 oz) per steak. This needs to be an intensely rich beef broth. You can make it from the trim of primal cuts by simmering the meat in beef stock, skimming off fat, and then adding more browned meat and vegetable scraps and concentrating it down again. You can substitute Demi-glace (see the YouTube video and notes in Volume 1, page 260). If you use Demi-glace, you'll need to add half a teaspoon of salt per 60ml (2 oz). As a last resort you can substitute Knorr beef gel packs (not dry cubes!) by dissolving one in 240ml (8.5 oz) hot water. That's enough for four steaks. It isn't as rich and it is a bit too salty, but it will get the job done in a pinch.

FOOD SAFETY

Some people have expressed concern about driving bacteria into the center of the meat with the Jaccard device. This is not a problem if you follow this method exactly as it is written. The bacteria won't multiply above 45°C (115°F) and the meat will be above that temperature shortly after you tenderize it. You are going to cook it so the internal temperature is safe anyway, so the only way there would be a problem is if you let it sit around uncooked at room temperature, and that's why you do Step #1 first and not second. Also, in the video I stress killing bacteria on the outside of the meat, but the key is to raise the *internal* temperature to at least 45°C.

PROCEDURE

1. Let the meat warm to room temperature for 30-40 minutes.

2. Use the single row Jaccard device along the sides of the meat, then the triple row on the flat face surfaces. You can use the single row on both, if that's all you have.

3. Season one side only of the meat with a little coarse salt (remember the broth is also salty, so don't use much) and fresh ground coarse pepper. Rub with olive oil.

4. Heat a cast iron (don't use nonstick here - stainless is okay, though) on very high heat (#10 out of 10) until it is smoking (at least 250°C / 480°F). Brown the steaks one at a time (don't crowd the pan) with the seasoned side-down.

5. Turn the steak over and give about 1 1/2 minutes of cooking time to the reverse and edges of the meat in all. You just need it to not be pink. Repeat for the other steak(s).

6. Crush the garlic and herbs in a mortar. Divide the herbs and garlic between the steaks. Ideally you want to use a vacuum sealing device, but you can use a plain plastic bag and suck the air out with a straw. Into each bag put a steak, the garlic and herbs, and 60ml (2 oz) of beef broth.

7. Use the largest stock pot you have for the water bath. A large volume

of water maintains its temperature for a long time and changes only slowly. Put an inverted steamer basket on the bottom to keep the bags of meat from touching the hot surface close to the stove. Put it on a burner and fill the pot, blending hot and cold water until it is filled and 52°C (125°F). You want to maintain the temperature at 50°C (122°F), but when you put the meat in, the temperature in the water will go down a little so you start it a bit hotter. Now put the bags in and set the stove heat to low. You'll have to experiment the first time to know what stove setting will maintain the temperature at 50°C.. Agitate occasionally for 2 hours. Keep a lid on the pot to help keep the temperature constant.

8. When the time is up, remove the meat and strain the contents of the bag through a sieve. Discard the garlic and herbs. Put the liquid in a pan on the stove on a medium heat (#5 out of 10). Bring to a simmer.

9. While the liquid is reducing, heat the cast iron pan once again to at least 250°C / 480°F and give each steak a good crust on both sides.

10. Put a pat of butter on top of each steak. They are ready now if you want them served rare. If you want them cooked more, transfer them to a 170°C / 330° F preheated oven with the pat of butter on top. Allow 7-12 minutes, depending on how well done you want them.

11. After 15-20 minutes, the sauce should be reduced. Put a tablespoon of butter in a mixing bowl for every two steaks you cooked. Pass the reduced liquid through a sieve into the bowl and whisk with the butter. Add a couple of drops of red food coloring, if desired.

12. Put some of the sauce down on each plate, then trim and slice the steaks. You can choose to trim the tough band that runs around each steak, or just slice it diagonally. The former will be easier to eat but the latter will look better on the plate. See the video for more about this choice.

RUTABAGA IN PORTER ALE
GARNISH FOR STEAK OR ROAST BEEF

As explained in the video, this is not intended to be a side dish. It is a plating element that is best used in a complex presentation where you only get a few pieces with other things.

180g (6.3 oz)	Rutabaga (also known as Yellow Turnips)
90ml (3.1 oz)	Porter Beer
30g (1 oz)	Shallots, sliced
30g (1 oz)	Butter (in all)
3/4 t	Salt
Sprig Rosemary (optional)	

PROCEDURE

1. Peel and cut the rutabaga into need cubes about 1.25cm (half inch) on each side. Then either blanch them in salted water for 30 seconds if you want them to keep their shape, or boil them for 8 minutes. The latter will result in a mushy texture, but better flavor in the end.

2. Drain and cool the pieces.

3. Heat a nonstick pan on a medium-high heat (#7 out of 10) and add about half of the butter. Add the turnip pieces to the butter and don't stir.

4. After 5 minutes there should be some good color on one said. Add the sliced shallots and the salt. You can stir it some now, but not too much.

5. After 3 more minutes add the porter beer. Reduce heat slightly. Cook until the porter has been absorbed. Add the rest of the butter and the rosemary. As it dries out, reduce the heat to medium-low (#4). Add a little black pepper. This is best when refrigerated and reheated the next day.

ROAST DUCK LEGS
WITH HAZELNUT STUFFING

An Austrian Christmas tradition developed after working with Wolfgang Puck during holidays many years ago.

700g (1.5 lbs)	Duck Legs, large
70g (2.5 oz)	Hazelnuts, peeled
2 T + 1 teaspoon	Austrian Seasoning Blend (page 221)
120g (4 oz)	Onions
60g (2 oz)	Celery
30g (1 oz)	Green Bell Pepper
3-4 cloves	Garlic, peeled
90g (3 oz)	Bread, fresh
1 whole	Egg
45ml (1.5 oz)	White Wine, dry (optional)
12 leaves	Sage, fresh
1 whole	Pear, fresh
1 T	Lemon Juice, fresh

PROCEDURE

1. Rub the seasoning into the duck legs. Use the tip of a knife to poke holes in the skin so that fat can drain during cooking. Allow to stand for 30 minutes to an hour at room temperature before you begin roasting.

2. During this time, peel the tough outer threads from the celery with a vegetable peeler. Then coarsely chop the celery, onion and green pepper.

3. Arrange the vegetables on the bottom of a baking dish that is the minimum size to hold the duck legs. The duck should cover almost all of the vegetables. If the ends of the legs stick out over the edge of the baking dish a little, that's okay because there's no meat at the tips of the joints.

4. Preheat oven. Pour the white wine into the baking dish if you are using it. If not, then add the same amount of water.

5. Roast at 160°C / 320°F for just under 2 hours. No fan assist.

6. Now add the garlic cloves to the baking dish and increase the oven temperature to 220°C / 450°F with fan assist ON. Return the dish with the duck to the oven for the final 15-20 minutes.

7. Reduce the oven temperature to 180°C / 355°F so that it has time to cool back down before it is needed again in step 12. Set the duck aside. Pass the juices and vegetables through a sieve, pressing down to extract as much of the liquid as you can. Use a fat separating pitcher to skim the fat. Save the fat for cooking the hazelnuts. Also reserve the vegetables that were strained off in the sieve.

8. Toast the hazelnuts in a hot dry pan for 2-3 minutes, then add a tablespoon or so of the duck fat and fry them until they are turning brown.

9. Add the cooked vegetables from the sieve in step 7 to the pan and turn off the heat. Cook for a minute, then transfer to a bowl. Allow it to cool.

10. Using a food processor, chop the mixture up so the nuts are coarse meal. Don't purée the mixture, though! See video if this is not clear.

11. Cut the bread into 1.25cm (1/2 inch) cubes. Put this in a bowl and add the contents of the food processor to it. Add the teaspoon of Austrian Spice Mix. Whisk the egg briefly and stir it into the same mixture.

12. Spoon this into the same casserole dish that the duck was cooked in. Roast at 180°C / 355°F for 40 minutes with no fan assist.

13. Loosen and "fluff up" the stuffing with a fork, but leave it rest in the hot baking dish. If you don't separate it and move it around, it will settle and become soggy. The steam trapped in it needs to vent.

14. Pan fry slices of pear in butter. Add lemon juice to balance the sugar.

15. When it is time to serve this, finish the duck legs by deep frying them at 175°C / 350°F oil for about 1 1/2 minutes. Then increase the oil temperature to 200°C / 390°F and deep fry some sage leaves as an edible garnish. A classic accompaniment to this is *Sweet and Sour Red Cabbage* (page 170).

SWEET AND SOUR RED CABBAGE

This dish is a gastronomic anomaly because it is widely consumed across most of Europe—for many people almost every day with meals. Yet it is almost never seen in the United States either at home or in restaurants. That's unfortunate because it is delicious and much healthier than, say, French fries. This version is a fusion of Swedish, Russian and Austrian styles, integrating the best of all three in my judgement.

360-400g (13.5 oz)	Red Cabbage, sliced (remove root stem)
70g (2.5 oz)	Brown Sugar, ideally Cassonade
30ml (1 oz)	Red Wine Vinegar
50-60g (2 oz)	Red Onion, sliced and chopped
30g (1 oz)	Butter
30g (1 oz)	Bacon Fat, rendered (see note below)
4-5 cloves	Garlic, sliced
1 teaspoon	Powdered Skim Milk (optional)
2 T	Austrian Spice Blend (see previous page)
to taste	Lemon Juice, fresh

BACON FAT

To make this more in the Swedish style, begin by cooking some bacon in order to get the rendered fat necessary. Then remove the bacon and set aside. Add it back in at the end of the cooking. Alternatively, just use some rendered bacon fat in the recipe, which is typical of Austrian cookery. Last, but not least, you can just use another 30 grams (1 oz) of butter.

PROCEDURE

1. Cook the butter by itself in a small stock pot on a medium heat setting (#5 out of 10). This is to make *Beurre Noisette* (browned butter that has a

natural scent similar to hazelnuts). When the butter melts, add the powdered skim milk if you are using it. This will intensify the flavor.

2. If you are using bacon fat as well, then add that to the cooked butter now along with the sliced garlic. Turn the heat down to low (#2 out of 10).

3. As soon as the garlic has some color on it, add the brown sugar. Stir.

4. After 2-3 minutes, increase the heat to medium (#5 out of 10) and add the cabbage. Fry it in the butter for about 3 minutes.

5. Now add the red onion. It should be cut to resemble the cabbage so that it blends in. Also add the Austrian Spice Mix. Cook for another 3 minutes.

6. Now add the vinegar. Stir and put a lid on the pan. Turn the heat back down to low (#2 out of 10). Cook for 1 1/2 hours. Stir every 15 minutes during this cooking time, replacing the lid after each time you stir it.

7. Remove the lid and increase the heat back to medium (#5 out of 10). At this point you want to drive off excess moisture and any off putting acidic vapors. You are also caramelizing the cabbage a bit. This is the stage that separates the professionally made product from the amateur home cook, so take care to not let it burn, but cook it well.

8. When it is close to dryness, taste it and adjust the flavor with more salt and some lemon juice (typically about a tablespoon of lemon juice.)

9. Transfer it to a bowl to cool at room temperature for 30 minutes before you refrigerate it. Although you can eat it right away, the best flavor is obtained by storing it in the refrigerator for 1 to 3 days first and then reheating it just before service.

COOKING BEETS
FOR THOSE WHO HATE BEETS

This is a method for preparing beets that will eliminate any foul odor or mineral taste that may be present.

2 whole	Beets, large
2 liters (8.5 cups)	Water
1 T	Salt
2 teaspoons	Red Wine Vinegar
2 branches	Rosemary, fresh
2-3 strips	Orange Peel (optional)
1/2 teaspoon	Cloves (optional)

THE BEETS

The taste depends on where they were grown and what sort of fertilizer was used. The skin is where most of the dirt and fertilizer taste is, so if you boil them with the skin on, you are driving that taste into the beets.

PROCEDURE

1. Rinse the beets off and then peel them. Now rinse them again.

2. In a large cooking pot, add the water and the vinegar, but not the salt. Put the peeled beets in. If necessary, add more water to cover the beets.

3. Bring to a boil, then partially cover and keep it at a boil for 45 minutes.

4. Add the salt and rosemary. You can also add some orange peel and a few cloves. Reduce heat to simmer for 30 minutes, still partially covered.

5. Now put the cover on completely and remove from the heat. Allow it to stand and cool for 30 to 90 minutes.

6. Remove from the water and refrigerate. Ideally wait until the next day.

+

ROASTED CAULIFLOWER
SCENTED WITH MUSTARD AND TURMERIC

The combination of cauliflower and mustard is a natural because they are closely related botanically, and share some of the same molecular flavor components.

1 whole	Cauliflower, small to medium head
30ml	Vegetable Oil
1/2 t	Dry Mustard (optional - see below)
1/2 t	Turmeric (optional - see below)
1-2 t	Lemon Juice, fresh
Sea Salt	

ROASTING CAULIFLOWER

While most people steam or boil cauliflower, this is a mistake because it absorbs too much water and the natural flavor in the cauliflower is washed into the water, leaving it very bland and soggy tasting. Dry roasting is a much better method.

PROCEDURE

1. Cut the cauliflower into thick slices so that you have flat surfaces. If this is not clear, then watch the video segment between 1:20 and 2:00.

2. Combine the vegetable oil, dry mustard, turmeric and lemon juice. Use a brush to paint the cauliflower slices. Turn them over and paint the other side. Now flip the pieces back to the first side and sprinkle with sea salt.

3. Roast in a preheated oven at 180°C / 360°F for one hour. Check the pieces after 45 minutes. If they are starting to singe, turn them over.

4. For best results, refrigerate and then reheat by frying in butter.

VEAL SCALLOPINI

The toasted rye flour adds a new dimension to this classic.

200g (7 oz)	Veal, thinly sliced
60ml (2 oz)	White Wine, dry
30ml (1 oz)	Madeira Wine
1 whole	Shallot
2 cloves	Garlic, chopped ahead of time
3 T	Parsley, fresh
2 T	Basil, fresh
to taste	Mushrooms (see note below)
30g (1 oz)	Butter
30g (1 oz)	Olive Oil
Rye Flour, additional Parsley, Lemon for serving.	

THE MUSHROOMS

Ideally you want to use fresh Porcini mushrooms for this, but that's probably not going to be an option. Don't attempt to use frozen or reconstituted dried Porcini for fresh. The next best thing is using a combination of Porcini mushroom paste and regular button mushrooms. You can purchase the paste online, or if you live where there are Porcini growing wild seasonally, then cook them and freeze them yourself when they are available. Failing both of these options, you can use a Maggi mushroom stock cube with the button mushrooms. The Maggi mushroom cubes are made from wild mushrooms and are very economical.

PROCEDURE

1. In a stainless steel pan (do not use nonstick for this because the temperature climbs too high for too long), add enough rye flour to cover the bottom of the pan to a depth of about 5mm (1/5th inch). Toast it with occasional stirring. Keep the stove heat to medium (#5 out of 10). After about 15 minutes the temperature will begin increasing rapidly and it will

begin darkening. Get it dark without actually burning it. Pay attention to the aroma and you'll see that at some point after about 20 minutes, even with regular stirring, you start to smell something burning. Pour it out into a tray immediately at that point to stop it from going any further. You might need to practice this a couple of times to get it right. I suggest you watch the video first to see how it should look.

2. Heat 30 grams (1 oz) butter over a medium heat. Slice the shallot and add it to the butter along with some of the mushroom paste or dried mushrooms that you have already reconstituted in hot water, if you are not using fresh Porcini, that is. How much you add is up to you, as explained in the video—it is a choice between aesthetics and flavor.

3. After about 3 minutes, add the garlic. Cook for another minute.

4. Now add the white wine, basil and parsley. Cook for about 2 minutes.

5. Add the Madeira. Then transfer to a stick blender cup and purée.

6. Pound the veal between sheets of cling film until it is thin and even.

7. Season both sides with salt and freshly ground black pepper. Then dredge in the toasted rye flour. Leave the pieces to rest a few minutes.

8. Trim off the stem end of the mushrooms and then slice them. Don't slice them too thin, though. They will shrink during cooking, remember.

9. Heat a large nonstick pan on a high setting (#8 out of 10). Add the olive oil. Get the temperature to about 180°C (365°F) before you add the veal in a single layer. Do not crowd the pan. Cook 2 minutes, then turn them over.

10. After about another minute or so, remove the veal to a plate. Next add the mushrooms to the pan with the hot oil. You can pour off some of the oil first if there is a lot, or if you are trying to keep the fat content down.

11. As the veal cools, juices collect. Add those to the mushrooms, too. When the mushrooms are brown, add the puréed sauce with the Madeira.

12. After a minute, add the veal back to the pan to warm through and finish cooking. Garnish with fresh parsley and a wedge of lemon.

FONDANT POTATOES
STOVETOP METHOD

This is a classic dish in French cuisine, however normally most of the cooking takes place in the oven instead of on top of the stove, as I am doing here. This is important if you are going to serve these with steak or roast beef, because the oven temperature for classic Fondant Potatoes is much higher than you can use for the meat, and most stoves only have one oven.

2 medium	Potatoes, in 20mm thick slices (0.8 inches)
45g (1.5 oz)	Butter (for amount shown)
1 T	Lamb, Beef or Pork Fat
	or just another tablespoon of butter
1	Beef Stock cube, ideally Gallina Blanca
3-4 branches	Thyme, fresh (optional)

ADDITIONAL NOTES

Before you cringe at the use of a stock cube in this classic dish, know that it is basically being used as flavored salt here. Also, Gallina Blanca and Knorr are nothing like the horrible bouillon cubes you may have tasted long ago. The technology has improved greatly, at least in too brands,

There is no consensus about the type of potato that should be used. There are fierce debates at both ends of the spectrum, with some chefs insisting on starchy potatoes and others claiming waxy is the key. For more about this, see page 19. I conclude that almost any potato will work, but the end result you get will vary accordingly. Just be sure to use a robust potato that won't crumble easily.

176

PROCEDURE

1. Select a nonstick pan that is just big enough to hold all of the sliced potatoes in a single layer. Don't put the potatoes in yet, though. Set the pan on a medium heat (#5 out of 10) to melt the butter and the fat in it.

2. When the butter has foamed up, add the sliced potatoes. Place a splatter guard over the top of the pan, if you have one. Cook for 5 minutes.

3. Now turn the potato pieces over. There is almost no browning yet, but that's expected. Reduce the heat a bit (#4 out of 10). Now cook on the other side for 5 more minutes.

4. Flip the potatoes back to the first side and reduce the heat a bit more again (#3 out of 10).

5. Crush the stock cube with a mortar and pestle. Sprinkle the top surface of each potato with some of the powder. Try not to get it on the pan, but only on the potatoes. Continue cooking at this low heat setting for about 20 minutes.

6. Turn the potato pieces over now. There should be deep browning on the other side (but no burning at all). Add the sprigs of thyme (if you are using them) and then reduce the heat further (#2 out of 10). Cook for 5 minutes.

7. Now turn off the heat. Flip the potatoes over again, coating them in the thyme-scented butter and making sure that the browned side is softened by soaking in the hot butter. Let them soak for 20 minutes, if possible.

SERVING

This is a classic accompaniment to steak, but they can be served as a side dish with many things. You can spoon some of the butter over the top and add a couple more sprigs of fresh thyme. Classically the cooking liquid would be discarded and the potatoes would be served after draining on paper towels, however normally the cooking liquid is mostly stock. These have more common ground with fried potatoes.

RED WINE SAUCE
A TYPE OF BORDELAISE

Even though this is an abbreviated version that enables you to make it just a couple of hours, the taste is far better than anything you will have in almost any restaurant these days.

1 bottle (750ml)	Red Wine, dry - ideally a Syrah (or Shiraz)
200g (7 oz)	Onion
100g (3.5 oz)	Carrot
100g (3.5 oz)	Celery
1 clove	Garlic, chopped coarsely
2 whole	Bay Leaves
2 t	French Herbs from Lyon (Volume 2, page 194)
1 T	Burgundy Spice Blend (page 230)
300ml (10.6 oz)	Beef Broth (see note below)
30g (1 oz)	Deluxe Italian Tomato Paste (page 122)
20g (0.7 oz)	Butter, cold

EMERGENCY SUBSTITUTIONS

You can use the Italian tomato paste that comes in a squeeze tube instead of the Deluxe Italian Tomato Paste. In an emergency, substitute water and a Knorr beef gel pack plus two sheets of gelatine (softened in cold water first) for the beef broth. As for the Burgundy Spice Blend, you should make that up ahead of time and keep it in your pantry (along with at least several of those other tertiary seasonings) to have on hand at all times. That little chemistry set will enable you to improve almost any recipe dramatically once you get the hang of what to use and when.

PROCEDURE

1. Peel and chop the onion, carrot and celery into small pieces. Put them

n a sauce pan and add half of the wine, the bay leaves, and dry herbs.

2. Bring to a simmer and reduce by half.

3. Add the beef stock and garlic clove, Return to a simmer and reduce by half again.

4. Add the rest of the wine, the Burgundy Spice Mix and the tomato paste. Return to a simmer and reduce until thickened.

5. Pass through a strainer and discard solids. Now pass through a very fine mesh strainer to remove any remaining bits.

6. Return to the stove and bring to a slow simmer. Adjust salt to taste. Remove from the heat and whisk in the cold butter until it is completely homogenous and glossy.

7. After first cooling to room temperature, store in the refrigerator. This will keep for up to two weeks in the refrigerator, but is best used within seven days.

"Reduce wine in pan by half,"—
I believe you can do it by turning up the heat.

HAZELNUSSOMELETTEN
GERMAN HAZELNUT OMELETTE

There are quite a few identical recipes for this online. It seems everyone copied the original one posted, and ironically that recipe was not especially good. This one is much better, and an example of how the Beautiful World seasoning mix can be used in dishes where you wouldn't expect it to work.

90ml (3.2 oz)	Cream, 22% fat
22g (0.7 oz)	Butter
1 whole	Egg, separated
2 T	Hazelnuts, grated (see text below)
2 T	Oat Flour
1 teaspoon	Sugar
1 teaspoon	Beautiful World Spice (page 233) - optional

GRATED HAZELNUTS

You will need about six peeled hazelnuts. Grate them on a fine microplane grater, discarding the nubs that you can't pass through.

PROCEDURE

1. Separate the egg into two bowls. Into the bowl containing the yolk, add the grated hazelnuts, oat flour and cream. Whisk together and let stand while you continue.

2. Heat a 22cm (8.5 inch) nonstick skillet on a medium-high setting (#7 out of 10) and add the butter to the pan.

3. Add the sugar to the bowl with the egg white. Using a hand-held mixer, beat the white until it forms firm peaks.

4. Fold about half of the flour/cream mixture into the egg whites, then the

rest of it, taking care to preserve the lightness of the mixture.

5. By now the butter should have foamed up in the skillet. If you have an IR thermometer, you want to see a temperature around 150°C (300°F). Use a spatula to scrape the contents of the bowl into the hot pan. Move it gently so that it covers the entire bottom surface. Lower the heat slightly.

6. Cook for about 5 minutes. During this time, dust the surface with the Beautiful World Spice Blend as evenly as possible, if you are using it.

7. When it is nicely golden with some brown patches (not burnt), flip it over using a large wide spatula. Cook on the other side for 2-3 minutes until it is also golden brown.

8. Transfer to a plate without flipping it over (the first side that was cooked in the butter should still face up).

9. Serve along with stewed fruit, especially apples and pears. Ideally you will stew the fruit yourself and add a touch of cognac toward the end of the cooking. Keep the effect almost subliminal in the background. Perfect for brunch, especially with champagne or mimosas.

FOOTNOTE

Hazelnussomeletten is almost unheard of outside of Germany and Austria. In an episode of the animated television series, *American Dad*, the talking German fish (Klaus) claims that Hazelnussomeletten has to be made with veal or it isn't authentic. This is completely false, of course. They counted on the audience not knowing what this dish is. The program is also aired in Germany, though.

MAFALDE WITH ROASTED CAULIFLOWER

As explained earlier (page 58), this recipe takes advantage of the reactive nature of the pink peppercorns to create an original flavor with the cauliflower.

450g (16 oz)	Cauliflower, trimmed
200g (7 oz)	Mafalde Pasta (or other)
150g (5.3 oz)	Onion, sliced
25ml (0.9 oz)	Olive Oil, extra-virgin
2-3 T	Basil, fresh
2-3 cloves	Garlic, coarsly chopped
1 1/2 teaspoons	Pink Peppercorns
1 teaspoon	Coarse Salt
2 teaspoons	Sicilian Spice (page 226) or a few Anchovies, rinsed
1/2 teaspoon	Baking Soda (not baking powder)
1/4 teaspoon	Nutmeg (optional)
Parmigiano-Reggiano	
Additional Olive Oil (see note below)	

THE ADDITIONAL OLIVE OIL

The second infusion of olive oil in this recipe can be recycled from the Marinated Broccoli Stems (page 142), or other high-quality Italian products that were packed in olive oil, if desired.

PROCEDURE

1. PREPARE THE CAULIFLOWER. Begin by grinding the pink peppercorns to a powder in a mortar and pestle.

2. Add the chopped garlic, the basil and the coarse salt. Also add the nutmeg at this point if you are using it. The flavor with the nutmeg is different, even though there is only a small amount of it—so I encourage

you to experiment and see which version you prefer. Now grind these ingredients with the pink peppercorns.

3. When it is ground to almost a paste, add the 25ml of olive oil and the baking soda. Grind more. Let the mixture to stand while you do step 4.

4. Cut the cauliflower into florets just a little larger than bite size (they will shrink some during cooking and also break up some. Discard leaves and tough stem parts (or save for another application).

5. Put cauliflower in a bowl. Add seasoning from the mortar and mix..

6. Spread the cauliflower out on a silicone mat set on a baking tray. The silicone mat is not absolutely necessary, but it will produce better results.

7. Roast at 190°C / 375°F for 15-18 minutes. No fan assist. Let it cool to room temperature. This part of the recipe can be done several days ahead.

8. Cook the Mafalde pasta in boiling salted water until just short of al dente. Break the pieces in half first, if desired. This makes it easier to eat.

9. In a large nonstick pan, heat 30ml (1 ounce) of the "additional olive oil" (see previous notes). Slowly brown the sliced onions in this oil.

10. After about 20 minutes when the onions are nicely browned, add the cauliflower. Stir to combine, then add the Sicilian Spice Blend, or the anchovies torn into small pieces. Cook for about 2 minutes.

11. Add the drained pasta to the pan. Mix together with tongs. Add a little more olive oil. Consider adding some other ingredients, too...

OTHER OPTIONAL INGREDIENTS

Prosciutto di Parma
Sundried Tomatoes
Marinated Artichoke Hearts
Marinated Broccoli Stems (page 142)
Caper Berries
Additional Basil

---

12. Transfer to a plate and shave Parmigiano-Reggiano over the top.

CHICKEN PROVENÇAL

Few, if any, regions of the world have gone through such radical changes as Provençe, France. Because of its strategic location, fertile soil and plentiful seafood, Provençe has been held by just about every powerful force in Europe going back thousands of years. As a result, the cuisine reflects a tapestry of styles and the detailed history of each dish is mostly forgotten.

1 kg (2.2 lbs)	Chicken Legs or bone-in, skin-on Thighs
2 T	Provençal Spice Blend (page 231)
2 T	French Provençal Herbs (in all)
	or combine the following dried...

	1 1/2 t	Summer Savory
	1 1/2 t	Thyme
	1 1/4 t	Marjoram
	1 t	Oregano
	1/2 t	Rosemary
	1/4 t	Sage

1/2	Lemon
100ml (3.5 oz)	Dry Vermouth, such as Cinzano *Extra Dry*
10 whole	Shallots
1 head	Garlic (whole cloves)

Olive Oil, Fresh Ground Black Pepper, MSG

FRENCH HERBS

Note that you need <u>both</u> the Provençal Seasoning and the French Herbs for this recipe. Also, if you are using a packaged herb mixture, be sure to check that savory is one of the main ingredients. This is critical to the flavor of this dish.

PROCEDURE

1. Put the flour and 1 tablespoon of the Provençal Seasoning Blend into a

arge bowl. Toss the chicken legs with the mixture, coating as evenly as possible.

2. Heat a large nonstick skillet on medium-high (#6-7 out of 10). Add enough olive oil to the pan to just coat the bottom after swirling. Fry half of the chicken legs until nicely golden on all sides. Use a splatter guard.

3. Remove the chicken legs to a large bowl. Now add a little more olive oil to the hot pan and fry the other half in the same way.

4. Transfer the rest of the chicken legs to the same bowl. Add the garlic and shallots to the hot pan. Turn down the heat and cook gently to soften them.

5. Add the other tablespoon of the Provençal Seasoning Blend to the bowl with the chicken legs as well as the 2 tablespoons of French Herbs. Toss to coat evenly.

6. Put the cooked garlic and shallots into a shallow ceramic baking dish. Add the vermouth, then the coated chicken pieces. Season with salt and fresh ground black pepper.

7. Remove any seeds in the lemon slices. Add the pieces of lemon between the chicken legs.

8. Roast in a preheated oven at 200°C / 390°F for 20 minutes. No fan assist.

9. Remove from the oven briefly to baste all of the chicken pieces with the juices in the pan. A turkey baster is convenient for this.

10. Return to the oven for another 20-25 minutes. No fan assist.

11. Add sprigs of fresh thyme. Also add a little MSG to each piece for an even more pronounced flavor.

SHRIMP GLAZED IN GUINNESS ALE WITH JALAPEÑO CHILIES

The deep mahogany color of these shrimp makes them really stand out. If you use the abbreviated version of the spice blend shown in the video they won't have the same color. These were originally part of a dish I called, "Nachos del Presidente" which featured freshly fried blue corn tortilla chips, a custom nachos sauce that had blue cheese in it, and Dos Equis beer from Mexico was used in place of the Guinness Ale.

300-450g	Shrimp, raw
120ml (4 oz)	Guinness Ale
30g (1 oz)	Red Onion, chopped
30g (1 oz)	Pickled Jalapeño Slices (see note below)
1 T	Texas Spice Blend (page 234)
2 teaspoons	Honey
1 teaspoon	Salt

Olive or Vegetable Oil, additional Pickled Jalapeños (optional)
Lemon Juice, fresh

ADDITIONAL NOTES

The bottled Jalapeño chilies are the pickled slices in jars that are usually used for nachos. Also, the Texas Spice called for here is far superior to the abbreviated version of seasoning used in the video.

PROCEDURE

1. Peel and devein the shrimp. Save the shells and tails.

2. Put a little vegetable oil in a hot saucepan and cook the shrimp shells and tails until they are pink and fragrant.

3. Add the red onion and salt. Continue cooking another 3 minutes on a

medium heat setting.

4. Add the Guinness and pickled jalapeño chilies and bring to a simmer. Cook for 6-7 minutes. The solution should be thicker now.

5. Transfer the contents to the cup of a stick blender and purée.

6. Pass through a fine mesh sieve, pressing to extract as much as you can. Discard the solids.

7. Add the honey and allow it to cool to room temperature. You can speed this up by putting it in the freezer, but don't actually chill it below room temperature because the marinade won't work well if it is cold.

8. Marinate the shrimp in the liquid for 30-40 minutes. Drain on paper towel. Use more towels to soak up moisture on the top of the shrimp. They don't need to be bone dry, but they won't cook well if they are wet, either. At this point you can store them in the refrigerator for up to 48 hours and fry them when you need them for service.

9. Either cook on a cast iron pan quickly, as in my French Bistro Shrimp recipe (see Volume 1 and the video), or better yet, cook over hot coals on skewers.

10. Add lemon juice and a little more salt. Serve with the following sauce:

60g (2 oz)	Mayonaisse
15ml (1/2 oz)	Lemon Juice, fresh
1 T	Texas Spice Blend (page 234)
1/2 teaspoon	MSG

CREAMY TEXAS MAYONNAISE SAUCE

Simply whisk together the above ingredients. This actually tastes better if it is made a day ahead of time and stored in the refrigerator for the flavors to develop, so plan ahead if you can. Put pickled Jalapeño chilies on the tails of the shrimp, as shown in the video.

BEEF AND NOODLE CASSEROLE

An American classic rooted in Hungarian flavors. Because this uses two complex preparations, you will need to plan this in advance. It also means that the result you get will be impossible for anyone to guess how you ever got such a complex flavor.

500g (17.6 oz)	Beef, bone-in short plate - lean
250ml (1 cup)	Beef Stock (or water + 1/2 Knorr gel pack)
100g (3.5 oz)	Mushrooms (see page 216 of Volume 2)
100g (3.5 oz)	Onions, peeled and chopped
50g (1.75 oz)	Carrots, peeled and cut into small pieces
50g (1.75 oz)	Celery, shaved of threads and diced
50ml (1.75 oz)	Brandy, ideally Veterano from Spain
60g (2 oz)	Puréed Tomatoes (passata)
6 cloves	Garlic, peeled and chopped
1 T	Hungarian Spice Blend (page 229)
150g (5.3 oz)	Tomatoes, fresh (cut into pieces)
60g (2 oz)	Smetana, 20-30% (or Crème Fraîche)
120g (4.25 oz)	Pasta, Cavatappi or Fusilli
3 T	Louisiana Crumble (page 245)
3/4 teaspoon	Paprika, Hot Hungarian (see page 36)
Vegetable Oil, Coarse Salt, Black Pepper, Fresh Herbs (optional)	

ADDITIONAL NOTES

There are a lot of ingredients and steps here, so be sure to know what you are going to be doing next at each step along the way to avoid being caught off guard, or leaving something out.

PROCEDURE

1. Season beef with coarse salt and freshly ground coarse black pepper. Brown well in vegetable oil on all sides. Transfer to a plate and hold.

2. Cut the mushrooms into quarters (or even smaller fractions if they are large). Fry in the same pan you just removed the beef from until they are

188

well browned. Add a bit more oil, if needed. Transfer to a plate and hold.

3. Add the carrots and celery to the same pan with more oil, if needed. Don't allow the oil to pool. You only want enough oil to keep things from sticking and not a drop more.

4. When the carrot pieces begin to show browning, add the onion and continue cooking for another 4-5 minutes, stirring occasionally.

5. Flambée with the brandy.

6. When the alcohol is burned off, add the passata, the garlic and the paprika (in that order). Cook another 3-4 minutes.

7. Add the beef stock and transfer the contents to a pressure cooker (assuming you didn't do all of this in a pressure cooker to begin with). Put the meat and the bay leaves in with the other ingredients and bring to a simmer.

8. Put the lid on and adjust heat to maintain pressure for 2 hours.

9. When the time is getting close, cook the pasta in boiling salted water until very *al dente* (about 2 minutes less than the package directions).

10. Release the pressure and remove the meat. Discard the bay leaves. Add the fresh tomatoes. Simmer the contents of the pan to reduce by about a third, skimming fat if there is enough to skim.

11. Now add the pasta to the reducing liquid and cook it until it is nearly soft enough to eat. Your goal here is to arrive at the point where the pasta is soft at the same time the sauce has thickened considerably. During this time you trim the meat and cut it into small pieces. Add this to the pasta.

12. Stir in the smetana. Transfer contents to a ceramic baking dish. Sprinkle on the Louisiana Spice Blend / Crumble. Bake at 180°C (350°F) for about 30 minutes.

13. Allow to cool for at least 20 minutes before eating. Even longer is better. Garnish with your choice of fresh herbs. Serve with a little additional smetana on the side.

TEA TOWEL CHICKEN BREASTS

Here is a foolproof easy method from the 19th century for cooking chicken breasts (either crowns or boneless/skinless) that are flavorful and tender with no skill required.

1	Chicken Crown, or 2 boneless/skinless breasts
2 cloves	Garlic, chopped
1 teaspoon	4-Mix Peppercorns (black, white, green, pink)
1/4 cup	Mix of fresh herbs (see notes below)
250ml (8.8 oz)	Chicken Stock (or water + 1 Knorr gel pack)

FRESH HERBS

Be sure to include fresh thyme. Other suggested herbs are rosemary, parsley, oregano, basil, celery leaves and dill. Do not use very much sage or marjoram (if any).

PROCEDURE

1. Grind peppercorns, garlic and coarse salt together in a mortar.

2. Tear the herbs into pieces with your fingers, adding them to the mortar. Grind some to bruise the herbs and release their flavor. Add the olive oil and continue grinding a little more.

3. If you are cooking a crown, then use the tip point of a sharp knife to poke holes in the skin. Otherwise ignore this step.

4. Spread the herb and garlic mixture over the chicken.

5. Fold in a tea towel. See video for how to do this. Place in a close-fitting oven-proof ceramic vessel. Pour the chicken stock over the towel. Put a lid on, or cover with two layers of foil.

6. Roast in a 210°C (410°F) oven. The time will depend on the texture you want. If you just want it tender, then an hour is enough. If you want it

oftest (especially useful for baked pasta dishes and ravioli fillings) then
et it cook for 2 hours, or even longer. If the chicken is still tough, put it
back for more. Eventually it will be tender, no matter what.

7. Take out of the oven and let stand at room temperature for 30 minutes
before transferring to a plate to unwrap.

8. Either peel the skin off of the crown and use the meat for pasta dishes
or other applications, or scrape off the herbs and place under a broiler to
crisp up the skin. Then rest, cut the meat from the ribs, slice and serve.

THE CURSE OF LOW FAT PROTEINS

Certain proteins present a difficult problem for cooking them. These all have one thing in common: Very little fat, because fat is moisture and tenderness. Examples include chicken breasts, lobster, octopus and shrimp. The worst of these by far are chicken breasts because there is a second problem, namely that chicken can't be served rare. In the case of lobster, octopus and shrimp, if they are a little undercooked that's not really a problem. So the solution is to cook them hot and fast and hope that the residual heat will perfectly finish them all the way through as they cool. Alternatively, with octopus it can be cooked for a long time to break down the proteins as much as possible. The worst case scenario for all of these proteins is the middle ground—where they are beyond the first sear, but not cooked nearly long enough to make them tender again. This approach with chicken is not the ideal method, I'll tell you right now. It was a 19th century solution that was kind of the *sous vide* approach of the day in which unskilled cooks could turn out something useable. The ideal method for cooking chicken breasts so that they are tender and flavorful means cooking them just to the point where they are barely safe to eat. Very barely. To do this reliably requires skill and practice that home cooks just don't have. That's the reason I don't show that way, because people will wind up getting sick trying to duplicate it. This method is easy and safe, but not perfect.

SOUTH TEXAS BBQ-STYLE BEEF RIBS

There are many different distinct styles of BBQ in Texas, and dedicated locals all claim that their method is the best and only "real' Texas BBQ. This recipe is original, but taken from the flavors of South Texas and the barbacoa of Northern Mexico. Braising the meat until tender before finishing it on a grill is considered blasphemy by Q-purists, but the method far more common in restaurants than most customers realize.

1.5kg (3.3 lbs)	Beef Short Ribs
100g (3.5 oz)	Red Onion, thick slices
60ml (2 oz)	Balsamic Vinegar
15g (0.5 oz)	Garlic cloves, peeled and halved
1 T	Texas BBQ Chili Powder (page 235)
1 1/2 teaspoons	Texas Spice Blend (page 234)
1 1/2 teaspoons	Chipotle Chili, dried (see note below)
1 1/2 teaspoons	Dark Brown Sugar, ideally Muscovado
1 teaspoon	Coarse Salt
1 teaspoon	Oregano, dried
3/4 teaspoon	Celery Leaf, dried
1/2 teaspoon	Black Peppercorns
1 whole	Bay Leaf, dried
<u>FOR THE SAUCE</u>	
60g (2 oz)	Ketchup
60g (2 oz)	Honey
15ml (1/2 oz)	Worcestershire Sauce
15ml (1/2 oz)	Liquid Smoke, ideally Hickory
2 teaspoons	Natural White Vinegar
1 teaspoon	Adobo Sauce (see note below)

ADDITIONAL NOTES

Note that the Texas Spice Blend is actually in this recipe twice, because it is also part of the Texas Chili Powder recipe. If you are using whole dried chipotle chilies, then first crush some in a mortar to measure out the powder. Dry the chipotles yourself, if possible (page 282). The adobo sauce called for here is the type that canned chipotle chilies come packed in. This will give it a pretty strong kick. If you are sensitive to spicy food, then reduce the amount or leave it out completely.

PROCEDURE

1. Grind the Texas Chili Powder and Texas Spice Blend together with the dried chipotle chili, brown sugar, salt, oregano, celery leaf, black peppercorns and bay leaf in an electric spice mill. Pour this over the ribs and massage in well.

2. Put the ribs into an ovenproof dish with a lid (suitable for braising) and then wedge the red onion slices between the ribs as spacers. Add the garlic cloves to the dish and put the lid on. Refrigerate at least 4 hours. A sort of crust will develop during this time. You can leave it overnight, too.

3. Remove the dish and let stand at room temperature for an hour before you start to cook it. When almost ready, preheat oven to 230°C (450°F).

4. Roast with no lid on for 30 minutes.

5. Drizzle the vinegar over the meat. Put the lid on and reduce the oven temperature to 150°C (300°F) for about 2 1/2 hours.

6. Allow to cool for 30 minutes, then remove the meat to a container. Pass the liquid through a sieve. Use a fat separating pitcher to isolate the broth.

7. Put the broth into a sauce pan along with the honey, worcestershire and vinegar. Simmer until reduced to about 170 grams (6 oz).

8. Stir in the adobo sauce and the liquid smoke. Refrigerate 1-2 days.

9. When you are ready to finish this up (up to 3 days later), either smoke the ribs indoors with a stove top smoker for about 40 minutes (control the amount of smoke by the amount of wood you add, and not the cooking time) — or, much better, cook over hot coals. Warm the sauce to re-liquify it. Glaze meat with the sauce and thin the rest with water to offer on the side. Onion rings are an excellent and side dish for this (see page 244).

CHICKEN SOFT TACOS & QUESADILLAS

This is another recipe that works well with the inexpensive kitchen vertical rotisserie, also known as a "Barbecue Maker" by some sellers. The Yucatan Spice Blend does an amazing job of filling in the background notes that you want in chicken quesadillas, and the tequila adds an extra dimension you won't find in very many restaurant versions.

650g (1.4 lbs)	Chicken (see note below)
30ml (1 oz)	Tequila, Añejo (or Mezcal)
30ml (1 oz)	Lime Juice, fresh
30ml (1 oz)	Corn Oil
30g (1 oz)	Onion, peeled
4-5 cloves	Garlic, peeled
3 tablespoons	Cilantro, fresh
30g (1 oz)	Jalapeño Chilies, pickled (from a jar)
2-3 whole	Dried Green Serrano Chilies
2 teaspoons	Yucatan Spice Blend (page 223)
2 teaspoons	Cumin Seeds
1 1/2 teaspoons	Oregano
1 1/2 teaspoons	Coarse Salt
1/4 teaspoon	Black Peppercorns
1/4 teaspoon	Cinnamon, ground

THE CHICKEN

The best results will be obtained with boneless, skinless thigh and leg meat. Breast meat will be drier, but still acceptable. Trim out any ligaments, either way.

PROCEDURE

1. In an electric spice mill, grind together the dried green chilies, cumin seeds, oregano, coarse salt and black peppercorns.

2. In the cup of a stick blender, put the tequila, lime juice, corn oil,

scallions, garlic, cilantro, fresh chilies, Yucatan Spice Blend, cinnamon, and the spices that were ground in step #1 above. Purée.

3. Cut chicken into pieces that can be threaded onto skewers. Put the chicken in a large bowl, taking care to remove any joints or bone fragments. Pour the marinade from step #2 over the chicken and massage it in. Refrigerate for at least two hours, or as long as overnight.

4. Thread onto the skewers of the vertical rotisserie, using a thick slice of carrot at the bottom to keep the pieces from sliding off. Roast for 15 minutes. Alternatively, cook over hot coals outdoors (even better). Finally, you can cook them 15cm (6 inches) from the heating element of a broiler.

5. Remove the skewers to cool for a few minutes, then use a fork to slide the meat off. Discard the carrot pieces.

6. Chop the meat coarsely and put into flour or corn tortillas. To make tacos, just add fresh salsa or pico de gallo and other condiments of your choosing such as grated cheese, sour cream, scallions, minced green chilies, guacamole and/or hot sauce and fold. For quesadillas, spread chopped chicken on a buttered flour tortilla on a nonstick pan. Add grated cheese, sprinkle with more Yucatan Spice and place another tortilla on top. Cook slowly to melt the cheese and brown the tortilla. Flip over and cook the other side. Serve with pico de gallo and sour cream on the side.

TOTALLY ALTERNATE VERSION

There is an American version of this in which you use apricot jam, Swiss cheese and plain cooked chicken. It's nothing like a traditional quesadilla, but it is interesting—and some people fall in love with this flavor combination, being reminiscent of the Monte Cristo sandwich (see my video on this and additional notes in Volume 1, page 253). If you sprinkle the Yucatan Spice Blend into the middle of the quesadilla, it will improve this version, too. It is like a kiddy meal at a family style Mexican-themed American restaurant, but there is certainly an audience for that.

✦

DOUBLE CHICKEN NORMANDY

The French name, Poulet à la Normande, is confusing because there are two classic dishes of the same name. These days we usually mean the one with chicken pieces braised with apples and cream. The other Poulet à la Normande refers to a whole chicken with a stuffing that includes chicken livers and Camembert. What I have done here is put the two recipes together.

CHICKEN & STUFFING INGREDIENTS	
1.6kg (3.5 lbs)	Chicken, whole
1 whole	Green Apple (divided)
70g (2.5 oz)	Chicken Livers (see note below)
45g (1.5 oz)	Camembert cheese
30g (1 oz)	Shallots
3 T	Parsley, fresh
1 1/2 teaspoon	Provençal Seasoning (optional)
1 teaspoon	Thyme, dried
25g (0.8 oz)	Olive Oil
60g (2.1 oz)	Bread Crumbs
90g (3 oz)	Ground Chicken or Pork
60ml (2.1 oz)	Cream
20ml (0.7 oz)	Vegetable Oil
15ml (0.5 oz)	Lemon Juice
1 clove	Garlic
1 T	Coarse Salt
2 t	4-Mix Peppercorns (black, white, green, pink)
1 t	Paprika
1 t	Thyme
1/3 teaspoon	Cloves, whole (the spice)
2 "petals"	Star Anise
150ml (5.3 oz)	Calvados (also known as Applejack)
1/4 teaspoon	Yellow Food Coloring (optional)

CHICKEN LIVERS

You can use just livers, but traditionally the stuffing also includes hearts and gizzards. Soak these parts in ice water for 5 minutes, then dry.

PROCEDURE – STAGE 1

1. MAKE THE STUFFING. Heat the olive oil in a skillet on a high heat #8 out of 10). Peel and coarsely chop half of the apple. Fry it in the olive oil for 5 minutes until lightly browned.

2. Coarsely chop the shallots and add those to the pan. Cook 2 minutes.

3. Add the chicken livers, hearts and/or gizzards. Cook for 2-3 minutes.

4. Cool near room temperature, then combine in a food processor with the Camembert cheese, bread crumbs, thyme, parsley, the Provençal Seasoning. Grind to coarse meal. Taste and adjust with salt and ground black pepper.

5. Put in a bowl and fold the ground meat in with a spatula, or by hand. Put in a plastic storage container and refrigerate for later use.

6. MAKE THE COATING. In an electric spice mill, grind the coarse salt, mixed peppercorns, cloves and star anise to a fine powder.

7. Put the ground spices into the cup of a stick blender along with the other half of the apple, oil, cream, lemon juice, paprika, thyme and garlic. Blend to a smooth purée. Add yellow food coloring now, if you like.

8. Rinse and pat the chicken dry. Place it on a roasting pan, breast side up. Using the tip point of a sharp knife, poke some holes in the skin, paying special attention to where the thighs meet the body.

9. Now put about half of the puréed mixture into the body cavity and smear it around. Put the stuffing inside and truss the chicken.

10. Apply the rest of the coating to the chicken as evenly as possible, using most of it on the top of the breasts and thighs. Dust lightly with flour and let stand for 15-20 minutes. Preheat oven to 180°C (355°F).

11. COOK THE CHICKEN. Roast for 1 hour 40 minutes in the preheated oven. After about an hour, baste the chicken with any juices that collected.

12. Transfer chicken to a platter and tent with foil to cool slowly. Deglaze the roasting dish with the Calvados. Note that you can substitute plain brandy or cognac for the Calvados, but the flavor will not be as good. Scrape up as much of the fond as possible, then pass through a sieve.

Reserve the liquid. When the chicken has cooled, section it and remove the stuffing. Save the carcass for making the sauce. Refrigerate the rest.

400ml (14,1 oz)	Apple Cider (the cloudy type)
2 whole	Green Apples, cored and sliced
120g (4.2 oz)	Onions, coarsely chopped
150ml (5.3 oz)	Cream
45g (1.5 oz)	Butter
1 T	Calvados
1 teaspoon	French Herbs, or dried Thyme

PROCEDURE – STAGE 2

13. MAKE THE SAUCE. In a large nonstick skillet, melt the butter on a medium-high setting (#7 out of 10). Whether or not you peel the apples is up to you. Cook the apples until they are golden. Remove the apples to a bowl and set aside.

14. Add the onion to the same hot pan the apples cooked in. There should still be plenty of fat (butter). Cook for 5 minutes.

15. Break up the chicken carcass and add it. Fry with occasional stirring for another 5 minutes.

15. Add the reserved Calvados that was used to deglaze the roasting dish and strained. Simmer to reduce by 75% (until thick).

16. Add the dried herbs and the apple cider. Note - in some countries the only apple cider you can get has alcohol in it. That will work just fine. Bring to a slow boil then turn the heat down. Simmer for 5-10 minutes.

17. Pass through a sieve. Discard the solids. Put the cream in a bowl and add a little of this liquid to the cream. Whisk as you gradually combine the two so that the cream doesn't curdle.

18. Return the liquid to the large nonstick skillet and return to a simmer once again. When it starts to thicken, add the previously cooked apples to

it. Preheat the oven to 220°C (450°F) with fan assist ON. Continue simmering for a few minutes to thicken. Remove the apples and set aside.

19. Pour the sauce into the bottom of a baking dish. Put the pieces of chicken on top of the sauce, skin-side up. Roast for 10-12 minutes. The skin should be slightly crisp.

20. Remove the chicken to a platter. Mix the apples back in with the sauce in the baking dish along with another tablespoon or so of Calvados. Divide the apples and sauce onto plates and add the chicken. Cut the stuffing into slices and warm. Garnish with a few sprigs of fresh thyme.

MORE ABOUT CHICKEN NORMANDY

This dish is a textbook example of how culinary trends work. Any recipe that is designated "Normandy" (and there are many in France) is taken to mean that it is cooked with a lot of cream and probably either apples or Calvados, because that region has a history of being rich in those ingredients going back more than a thousand years. Camembert is also a product of Normandy, dating back to the 18th century.

If you search American cookbooks published before the 1940's (*i.e.* World War 2), you will rarely see any meat dish cooked with fruit. When soldiers returned from Europe after the D-Day invasion at Normandy, they had been in France long enough to have tasted flavor combinations that were previously unknown to them. In particular, cream sauces which became synonymous in the minds of Americans with all of French cooking—a myth that continues to persist. Also the idea of cooking meat with apples, such as pork chops with applesauce.

The reason why so many American recipes call for mushrooms in this dish is due to the confusion between the two famous "Chicken Normandy" recipes. The French recipe with the Camembert stuffing is served with mushrooms. The French recipe with the apples and Calvados doesn't have mushrooms.

CHINESE XO SAUCE

In the same way that the complex tertiary seasonings at the back of this book (page 217) amplify flavors but are not the primary flavor in a dish, so too is the concept of XO Sauce. It can be added in a small amount to a great many dishes to kick them up a notch, to borrow a well known catch phrase.

200g (7 oz)	Scallops
60g (2 oz)	Shrimp, peeled (cooked or raw)
50g (1 3/4 oz)	Lightly Smoked Pork Salami
90ml (3.2 oz)	Peanut Oil (in all)
30ml (1 oz)	Sesame Oil
90g (3.2 oz)	Leek, white and tender light green parts
2 whole	Knorr Chicken Stock Cubes
25g (0.9 oz)	Garlic cloves
40g (1.4 oz)	Sugar
4-5 whole	Dried Thai Chilies
2 teaspoons	Soy Sauce
1 teaspoon	White Pepper, ground
1 teaspoon	Chili Oil

ADDITIONAL NOTES

In addition to substituting more readily available ingredients in a couple of places, I have made this less oily and not quite as spicy hot. This gives you more control when using it as a cooking ingredients, since you can always add more oil and chilies to something, which makes this all about the unique flavor profile of the caramelized seafood.

You can use a commercial chili oil, or the one that I provided the recipe for on page 102 of Volume 1.

Also, in my experience, you can use scallops and shrimp that have some freezer burn on them with perfectly fine results, making this an ideal use for seafood that would otherwise be inedible.

PROCEDURE

1. In a food processor, grind together the scallops, shrimp and salami.

2. Add 60ml (2 oz) of the peanut oil to the food processor bowl along with the sesame oil, leek (coarsely chop first), crumbled stock cubes and crumbled dried chilies. Process until almost a paste.

3. Heat the remaining 30ml (1 oz) peanut oil in a 3-4 liter stock pot on medium (#5 out of 10). When it is hot, add the contents of the food processor. Stir frequently to keep it from sticking and burning at first.

4. After a couple of minutes, the mixture will become soup-like, at which point you can stir it less frequently until it thickens up to an oatmeal-like stage. This will take about 15 minutes. Mince the garlic during this time.

5. When it is starting to stick to the bottom and has thickened noticeably, add the sugar. Reduce the heat slightly to #4 out of 10. Continue cooking and stirring another 5 minutes.

6. Now add the minced garlic and soy sauce. Continue cooking another 5 minutes with frequent stirring.

7. Turn the heat down to low (#2 out of 10) and add the white pepper and chili oil. Cook for 2-3 minutes more.

8. Transfer to a bowl to cool down to room temperature, then move to a glass jar to store in the refrigerator. It needs to chill for at least 24 hours to develop the flavor. This has a long shelf life as long as you keep it cold.

LOWER COST ALTERNATE VERSION

You can substitute 270 grams (9.5 ounces) of calimari for the scallops. Previously frozen calimari is just fine. The rest of the recipe is otherwise identical, except that stage 4 will take about 3 minutes longer. The taste is not the same, but it is similar and can be used in the same way. Most people won't notice it in a final cooked dish in which XO Sauce is only a minor ingredient.

XO CHICKEN HEARTS

The name XO Sauce comes from the designation on XO (extra-old) cognac, which is the highest commercially marketed type. XO is also an old way of writing a hug and a kiss in a romantic letter, so here we have hearts cooked with spicy red chilies and flambéed with cognac. You can use XO cognac if you have the budget, but it's an exercise in extravagance.

200g (7 oz)	Chicken Hearts, washed and dried
1 1/2 teaspoons	Yucatan Spice Mix (page 223)
1 teaspoon	Flour
1/2 teaspoon	Salt
25ml (0.9 oz)	Vegetable Oil
30ml (1 oz)	Cognac or Brandy
45g (1 1/2 oz)	Tomato Purée (passata)
2-3 cloves	Garlic
1-2	Red Serrano Chilies, fresh
45g (1 1/2 oz)	Chinese XO Sauce (page 200)
1	Scallion, fresh - green part only
1 sprig	Oregano, fresh (optional)

ORIGINAL EL SALAVADOR MODIFICATION

In the video, a simple spice blend is used in place of the complex Yucatan Spice Blend. That mixture was close to the way it was made by the chef who showed me how it was done in El Salvador. Only instead there were more dried chilies, regular paprika (not smoked) and fresh lemon zest. That's also good, but this version with the Yucatan Spice Mix and the XO Sauce is the winner by a long shot. The only real trick to this is in not overcooking the chicken hearts. Having confidence to cook them quickly and know they are done may take some practice.

PROCEDURE

1. Rinse the chicken hearts in cold water. Dry them and put them in a bowl. Season with salt. Then toss with the Yucatan seasoning mixture to coat evenly. Then add the flour and toss again to coat with that on top. Set the chicken hearts aside at room temperature for 15 minutes to an hour.

2. During this time dice the red chili. If you don't want it as hot, then scrape out the membrane and seeds. Also cut up the green scallion now and slice the garlic.

3. Heat a large nonstick skillet on a high heat (#8 1/2 out of 10). When it is hot, add the vegetable oil to the pan. Wait about a minute for the oil to come up to temperature and then add the chicken hearts. Sauté them for about 45 seconds.

4. Add the garlic and the chilies. Cook for 1 minute.

5. Add the cognac and flambée. Cook for about 45 seconds.

6. Add the tomato purée. Stir and cook for another 45 seconds.

7. Add the XO Sauce. You can use the lower cost version, but the flavor won't be quite as good (be realistic in your expectations when cutting corners). Now turn the heat off. Add half of the green scallions. Move the pan off the burner and cook with the residual heat for another minute or so. Don't overcook the chicken hearts! That's what makes them tough.

8. Transfer to a plate and then add the rest of the green scallions. A sprig of fresh oregano is a nice aromatic touch.

SZECHWAN CHILI PRAWNS

Largely famous as the signature dish of Iron Chef Chen Kenichi. Almost every recipe you see for this uses bottled chili sauce and ketchup. This is how to do it the right way.

200g (7 oz)	Large Shrimp, raw and unpeeled
3 whole	Scallions
1 teaspoon	Sesame Oil
22ml (0.75 oz)	Shaosing or Sherry wine
60g (2 oz)	Red Serrano Chilies
60g (2 oz)	Peanut Oil (in all)
30g (1 oz)	Bay Shrimp (see note below)
1 T + 1 t	Light Brown Sugar or Honey
25g (0.75 oz)	Garlic cloves
1 1/2 t	Corn Starch
60g (2 oz)	Tomato Purée
45g (1 1/2 oz)	Chinese XO Sauce (page 200)
1 T	Ginger, freshly grated
2 teaspoons	Lime Juice, fresh (optional)

BAY SHRIMP

Really you want to use 15 grams (0.5 oz) of *Hom ha* (Hong Kong Shrimp Paste). However this can be very hard to find. There are many different products from Asia that are translated as "shrimp paste" in the English language label. Many don't even contain shrimp! If you can't get Hom ha, then stick to bay shrimp.

PROCEDURE

1. Peel and devein the shrimp. Save the shells, but discard the "veins" (which is actually a digestive tract).

2. Put the shelled shrimp in a bowl and mix with the corn starch. Cover the bowl and put it in the refrigerator for later use in the recipe.

2. Heat a small sauce pan on medium-high (#7 out of 10). When it is hot,

put in the sesame oil and then the shrimp shells and tails.

3. Chop up the white part of the scallion. Save the green part for later use. After the shrimp are well cooked (about 4 minutes) add the white scallion.

4. After 2 minutes, add 150ml (5.3 oz) water and lower the heat to medium (#4 out of 10). Maintain at a simmer for 10 to 15 minutes.

5. During this time trim the red chilies. If you want to limit the heat, then scrape out the seeds and membranes. However, this is supposed to be a very hot dish with all of the heat that these chilies can bring. Suit yourself.

6. Put the chilies in the cup of a stick blender along with 45 grams (1.5 oz) of the peanut oil, the brown sugar (or honey), the garlic cloves and the bay shrimp (or Hom ha, as explained in the notes on the last page). Purée.

7. Set a nonstick skillet on a medium heat (#6 out of 10) and fry the puréed paste. You don't need any additional oil added to the pan.

8. While this is frying, rinse out the stick blender cup and pour the contents of the shrimp shells, white scallions and water into it. Add the Shaosing wine and purée. Pass this through a sieve and discard the solids.

9. When the oil begins to separate in the pan with the chili paste frying (about 7 minutes from when it began cooking), transfer the contents to a bowl Put it in the freezer for a few minutes to cool it down rapidly.

10. Once it is below room temperature (but not frozen - about 8 minutes) mix the chili paste in with the shrimp that were coated with corn starch.. Let them stand at room temperature for 10-25 minutes before continuing.

11. During this time, chop the green part of the scallions and grate the ginger. Then heat a very large nonstick skillet on high (#8 out of 10).

12. Add a tablespoon of peanut oil to the pan. When it is smoking hot, fry the shrimp and marinade mixture for just a few seconds over 1 minute.

13. Add the tomato purée and ginger. Cook for another 1 minute.

14. Add the stock made from the shrimp shells to cool it off and thin it. Remove the shrimp to a platter. Now add the XO Sauce and half of the chopped green scallions to the pan. Lower the heat some and reduce the sauce until thickened to your taste. Adjust the salt. Add rest of the scallions and a little lime juice.

✦

SUPER TAPENADE

There's a whole lot of molecular chemistry going on in this recipe, and like Chinese XO Sauce (page 200) it is something you really have to experience to understand. This isn't a tapenade in the usual sense. It is a flavor booster for black truffles and wild mushrooms. The initial taste is sweet. This begins to fade on your palate as you taste olives and black truffles. Then finally a deep umami taste builds with a long finish. A little of this can add some serious magic to many dishes, especially steaks and pasta.

90g (3.2 oz)	Red or Orange Bell Pepper
90g (3.2 oz)	Black Olives, ideally Kalamata
75ml (2.6 oz)	Red Wine, dry
60g (2 oz)	Tomato Purée (passata)
30ml (1 oz)	Olive Oil, extra-virgin
5-6 cloves	Garlic, sliced
1 1/2 teaspoons	Baking Soda (not baking powder)
1 whole	Knorr Beef Stock cube
2 tablespoons	Black Truffle shavings / trim (optional)

ADDITIONAL NOTES

At one point this is going to liberate a smell that's quite unpleasant. Don't worry. The smell will change within a few minutes.

PROCEDURE

1. Pit the olives.

2. Heat the olive oil on a medium heat. Cut the bell pepper into coarse dice. Cook the bell pepper for about 5 minutes to sweat it.

3. There should be liquid in the pan now from the pepper. Now add the garlic and cook until it is just barely starting to turn golden.

4. Sprinkle the baking soda on and stir. Cook for about 45 seconds.

5. Add the tomato purée and stir. Cook for about 2 minutes.

6. Add the wine. The mixture should turn almost completely ink black if you have done everything right. Stir to combine and lower the heat.

7. Powder the stock cube in a mortar and sprinkle into the mixture. Stir to combine.

8. Transfer to the cup of a stick blender. Add the olives and the black truffle scraps (if you have them) and purée. Store in the refrigerator in a glass jar. It has a long shelf life and actually improves with age. It is much better a week after you made it than when it is fresh.

STEAK WITH BLACK TRUFFLES

Brush a little on steaks going into the oven for roasting to give them a touch of black olive essence, which is ideal if you follow this up by topping the steak with shaved black truffles—the steak of kings!

THE ORIGIN OF SUPER TAPENADE

Despite the name, this is not a true tapenade. Sure, it has ground up black olives and some other common elements, but the chemical reactions that take place when you make this turn it into something that's only vaguely related to an actual tapenade.

Black truffles are one of the most costly ingredients for a restaurant. A long time ago I experimented with thinning out the black truffle pureé with an ordinary tapenade. As long as it was still about 90% truffle, no one could tell the difference and 10% savings is a lot of money when it comes to a high volume of truffles. Still, it wasn't in the spirit of a fine restaurant to cheat customers by watering down quality ingredients. The way to get around this was to make a tapenade that actually improved the flavor of the black truffle with molecular trickery that reinforced the natural flavor. This is the result after dozens of experiments.

CAPELLINI WITH GARLIC AND BUTTER

This version would be considered blasphemy by most Italians, but only if they find out how it's made. When it's served in restaurants, this dish receives only praise. There is molecular gastronomy afoot here, though it may not be obvious. See the history on the opposite page for more.

100g (3.5 oz)	Leeks, tender part only (do not substitute)
45ml (1.5 oz)	Olive Oil, extra-virgin
45ml (1.5 oz)	White Wine, very dry
130g (4.6 oz)	Cappelini (Angel Hair pasta)
30g (1 oz)	Butter
30g (1 oz)	Parmigiano-Reggiano, grated
8 cloves	Garlic, peeled and minced (in all)
1/2 teaspoon	Fennel Salt (page 21), or MSG
Parsley, freshly minced	

ADDITIONAL NOTES

Be sure to use real Italian Parmigiano-Reggiano. Also note that the garlic-leek purée (steps 1 & 2) is enough for three servings of the pasta, and the latter half of the ingredients list is for *each* of those three portions.

PROCEDURE

1. Roughly chop the leeks and combine them in a food processor with four of the garlic cloves. Grind as fine as possible (there will still be some larger pieces, and that's okay). Allow the mixture to stand 5 minutes.

2. Transfer to the cup of a stick blender. Add the olive oil and white wine. Purée. Note: Do <u>not</u> use the stick blender for step 1 or this won't work.

3. Begin cooking the pasta in lightly salted water. Don't boil it too fast.

208

4. Melt the butter in a large nonstick pan on a medium-high heat.

5. When the butter has foamed up, reduce the heat to medium (#5 out of 0) and add a tablespoon or so of the garlic. Cook 3-4 minutes.

6. Add about 50 grams (1.8 oz) of the purée from the stick blender cup. Cook gently for 4-5 minutes. You may need to lower the heat on the pasta or the sauce so that the pasta is al dente when the sauce is done.

7. Add the pasta to the pan and 1/3 of the parmesan. Toss to mix well.

8. Increase heat to medium again. Add another clove of minced garlic and another third of the cheese. Add a splash of the pasta water and toss.

9. A minute later add the fennel salt (or MSG) and turn the heat off. Toss.

10. Add a little more water if it is looking dry. Transfer to a serving plate and top with the rest of the cheese. Add freshly minced parsley. Serve with additional parmesan available.

HISTORY OF THIS DISH

I lived in Hollywood during most of the 1980's. There were a great many delivery restaurants at the time. Sometimes there would be over a dozen different flyers on the door and in the mailbox in a single day. One flyer always made me laugh because they claimed to have authentic Italian food, but the menu also had items like wonton soup and egg rolls on it. I showed a fellow cook the menu, expecting him to laugh. Instead he defended the place, saying I had to try it. Especially one dish in particular—this one. He convinced me. I wasn't madly impressed after the first bite, but the more I ate, the more I began to appreciate something special was going on. This wasn't just garlic and white wine. What was this? I tried repeatedly to duplicate it at home, but it was never the same. If you read my book, *40 Years in One Night*, you know how I would work at restaurants for a while just to learn their secrets. It wouldn't be until several years later that I realized what the chemical explanation was for this subtle difference in flavor. I covered the reaction earlier in this book (pages 78-79).

PENNE IN ARRABBIATA SAUCE

All too often this dish is made as if it was just some crushed pepper flakes added to a simple pomodora sauce, but this is a distinct sauce with it's own unique flavor when made properly. Although the name means "angry", this is still Old World cuisine, so the heat level shouldn't blow your head off.

320g (11.3 oz)	Italian Tomato Purée (passata)
310g (11 oz)	Penne Rigate
200g (7 oz)	Tomatoes, vine ripened
45g (1.5 oz)	Red Onion, coarsely chopped
15g (0.5 oz)	Garlic cloves, peeled and halved
1 whole	Dried Cayenne Chili Pepper (see below)
1 teaspoon	Crushed Red Pepper Flakes, hot
1 teaspoon	Coarse Salt
1-2 T	Basil Leaves, fresh (optional)
1 1/2 teaspoons	Sicilian Spice Blend (page 226)
	or 2-3 dried anchovies, as in the video
	or 1-2 fresh anchovies, rinsed

Extra-Virgin Olive Oil, Butter (optional)
Parsley, fresh; Pecorina Romana Cheese, freshly grated

DRIED CHILI PEPPERS

Ideally you want to include a whole dried cayenne chili that you grind up yourself, but this may not be possible depending on where you live. You can substitute a dried red Serrano chili, or if worst comes to worst, just use another teaspoon of crushed red pepper flakes.

PROCEDURE

1. Put the red onion and garlic cloves into a food processor and chop finely. Use a spatula to scrape the sides and make sure that the pieces are in contact with each other, but don't actually purée this! You need visible pieces for this to work.

2. In an electric spice mill, grind the dried cayenne chili with the coarse salt. Even if you are using Serrano chili or crushed red pepper flakes, do the same thing.

3. Heat a sauce pan on a medium heat (#4 out of 10). When it is warm, add enough olive oil to coat the bottom (about 50ml, or 1.7 oz). Add the ground cayenne and salt to the oil. Cook slowly with occasional stirring for 8-9 minutes.

4. Make a conical incision into the tomatoes to remove the stems, then cut them into about 8 pieces each.

5. When the olive oil is nicely colored and aromatic, add the crushed red pepper flakes to it and position the tomato chunks so that the skin-side is down on the pan. Increase the heat slightly to medium (#5 1/2 out of 10).

6. After about 6 minutes add the garlic and red onion from the food processor. Stir occasionally while cooking another 3 minutes.

7. Add the tomato purée and the Sicilian Spice Blend. If you have a stem from the tomatoes, then add that to the pan, too. Simmer for 8 minutes.

8. Add basil leaves and turn off the heat. Cook by residual heat 2 minutes.

9. Pass the mixture through the coarse disk of a food mill. This is enough for two batches of 155 grams (5.5 ounces) each of penne rigate pasta.

10. MAKE THE PASTA. Bring a pot of salted water to a boil and begin cooking the pasta. Heat a very large nonstick skillet on medium (#5 out of 10) with 2 tablespoons of butter (you can do half at a time if you prefer, but be aware that the spicy heat fades pretty fast - mostly gone in a day).

11. When the pasta is not quite al dente (typically 9 minutes), use a spider to remove it to the pan with the hot butter. Increase the heat to medium-high (#7) and fry the pasta in the butter for 3 minutes. As an option, add a little Hungarian Spice Blend (page 229) halfway through this phase.

12. Add the Arrabbiata Sauce and about 100ml (3.5 oz) of the pasta water. Cook until the sauce is absorbed and the pasta is ready to eat. Add a little more Arrabbiata. Stir, then plate with minced parsley and grated romana.

DEATH BY CHOCOLATE CAKE

I rarely mention desserts unless I have something worthwhile to contribute to the topic. I'm not a baker and I have a great respect for that distinctly separate craft. This cake is moist and delicious, but unorthodox. Bakers have some set rules for how to produce batters, and this doesn't follow those general rules, but it works if you follow the directions.

100g (3.5 oz)	Butter
2 whole	Eggs
80g (2.8 oz)	Cocoa Powder, 100% cacao
70g (2.5 oz)	Flour, all purpose
60g (2.1 oz)	Sugar, white
50ml (1.75 oz)	Cream, 22% fat
50ml (1.75 oz)	Milk
1 teaspoon	Baking Powder
1/2 teaspoon	Baking Soda
1/2 teaspoon	Vanilla Extract
1/4 teaspoon	Salt
5 squares	Lindt 85% Chocolate (optional)

PROCEDURE

1. Leave the butter and eggs out at room temperature for about half an hour before starting. The butter should be just slightly soft.

2. Put the softened butter, sugar and salt into the bowl of a stand mixer. Using the paddle attachment, cream the mixture.

3. When it is pale and light, stop the mixer and add the whole eggs, baking soda, baking powder and vanilla extract. Run the mixer with the paddle for another 2 minutes or so to blend these ingredients together, stopping to scrape down the sides after the first minute.

4. Now, with the mixer running slowly, add the cream. Then put in the

cocoa powder a little at a time. Then the flour.

5. Remove the paddle attachment. Scrape any batter from the paddle back into the bowl. Change the attachment to the whisk.

6. Run the mixer up to maximum speed and maintain it for 3 full minutes.

7. Preheat the oven to 170°C / 340°F. Butter a 6-place silicone cupcake mold where each cup is 8cm (3.25 inches).

8. Fill <u>five</u> of the cups about a third full with batter. Then place a square of chocolate on top of each. Divide the rest of the batter among the cups. Don't worry about them not being flat on top because they will melt and liquify as soon as they warm up in the oven, making them even for you.

10. Bake for 22-25 minutes. Remember that ovens vary some.

11. Let them cool for at least 10 minutes. There is some cracking on top, but it doesn't matter because you turn them upside down to plate. Sprinkle with powdered sugar. A dollop of whipped cream and some fresh berries complete this dessert. These can be reheated in a microwave for a few seconds with surprisingly good results.

CHOCOLATE CHIP COOKIES

By popular request, I am including another dessert recipe. This is for about 10 cookies. You can scale it up by simply multiplying the ingredients as many times as you need.

130g (4.6 oz)	Flour, all purpose
75g (2.6 oz)	Butter. slightly softened at room temperature
75g (2.6 oz)	Sugar, white
30g (1 oz)	Dark Brown Sugar, ideally Muscovado
1 T	Almond Flour (finely ground almonds)
1 T	Soy Flour
2 whole	Eggs
1 teaspoon	Vanilla Extract
3/4 teaspoon	Baking Powder (not baking soda)

Pecans or Macadamia Nuts, broken into pieces
Lindt Chocolate, 70% cacao

ALMOND AND SOY FLOUR

You can substitute additional ordinary all-purpose flour for these ingredients, but the flavor and texture will not be as good. The soy flour makes the cookies moister, so if you prefer dry, hard cookies (and some people do) rather than a slightly chewy texture, then definitely substitute regular flour for the soy flour. If you have a nut allergy, then you can make these without the almond flour or pecans, of course.

PROCEDURE

1. Cream the butter and sugar together in a stand mixer with the paddle attachment. Pause to scrape down the sides, as necessary.

2. Add the eggs, the soy flour, the vanilla and the baking powder. Power the mixer back on low to gently combine. Stop and scrape down the sides as necessary.

3. Running the mixer slowly, shuffle the flour in a little at a time to avoid
t being blown out and across the table. Again, pause and scrape down the
sides and scrape off the paddle with a spatula to keep the mixture together
in the bowl.

4. Refrigerate the dough for at least 1 hour. Up to 3 hours is okay.

5. Preheat the oven to 180°C (360°F). While the oven preheats, put a
silicone mat down on a baking sheet. Use a tablespoon to make little
mounds of the dough. You should have enough for 10 cookies of the
optimum size. Arrange like this, leaving plenty of space in between..

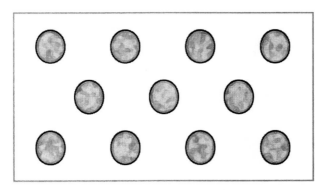

6. Now break the pecans into chunks and press lightly down into the top
of the batter. Don't mix them into the batter, or they won't toast. The fact
that the chocolate and nuts are entirely on top is part of what is good about
his recipe.

7. Now do the same thing with the chocolate. Don't make the pieces too
small. The cookies are best with some nice large globs of chocolate and
nuts sticking out.

8. Bake for 18 minutes.

9. Cool the cookies on a wire rack for at least 30 minutes, and a few hours
is arguably even better. If you are a perfectionist, or you just need these to
look picture perfect for a restaurant, use a ring mold to trim the irregular
edges while they are still warm.

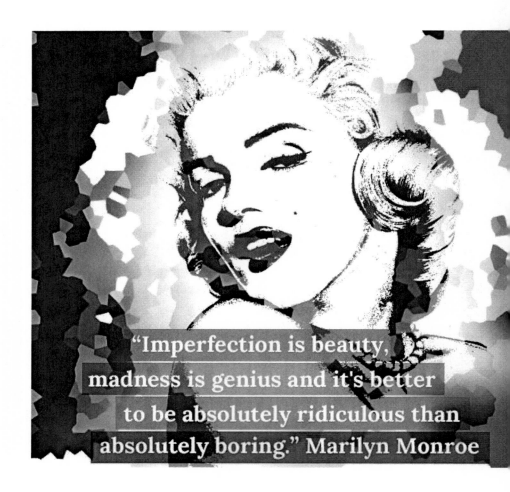

"Imperfection is beauty, madness is genius and it's better to be absolutely ridiculous than absolutely boring." Marilyn Monroe

* You'll understand why I included this quote after you read the next section.

Spices & Coatings

When you read the label of most commercial food products, you'll see the catchall phrase, "natural and artificial flavors" listed. We'll ignore artificial flavors, because you can't get the chemicals necessary and you don't want to use that stuff anyway. By the way, MSG is a natural flavor, but that's another topic. Many of the flavors that are being added these days are *tertiary*, but few consumers understand what that even means.

TERTIARY DEFINITION

Tertiary means the third in line. For example, the primary flavor of a fast food burger is the meat. The secondary flavors are the condiments and to some lesser extent, the bread. The tertiary flavors are notes that you can't consciously pick out, but they make foods taste fresher and more vibrant. These additives leave food tasting natural when used in moderation.

The industrial extraction of natural flavors (so that you can have a whole bottle of a specific additive) involves equipment that no home kitchen has, so my challenge was to find another means of accomplishing this. Toasted flour is my innovation (page 13) although it may strike you as ridiculous, it provides...

- *A substrate of fine particles to absorb and store flavors*
- *Diluting intensity so you aren't measuring in micro fractions*
- *To promote browning on surfaces*
- *To promote thickening in sauces*
- *As a flavor component directly (toasted flours have flavor)*

I named these seasonings after different parts of the world to help you remember what nationality the flavor is associated with. It's a guideline that usually works, but it isn't an absolute.

APPLICATION GUIDE

The easiest mistake to make with many of these seasonings is to try to use them as primary flavors. In some cases you might be able to come up with something good, but more often the results will be like that pirate back on page 133. In the same way you can't flavor a dish solely with MSG, yet MSG can add a dynamic boost to a savory dish when used *just below the threshold of conscious perception*. This is the key to "natural flavorings" you see on labels.

NOTE: DO NOT SUBSTITUTE DRIED PRODUCTS FOR FRESH! (*e.g.* raisins instead of grapes). The chemistry here relies on the exact compounds specified to react or to be absorbed on the flour.

SYMBOL	MEANING	APPLIES TO...
👍	In some cases it can actually work as a primary flavor, but still designed to be used as a small amount to augment other flavors.	Austrian Spice Blend Oaxacan Spice Blend Thai Spice Blend Worcester Spice Blend Provençal (France) Spice Blend Ottoman Spice Blend
💧	Needs fat or oil to function. Otherwise the flavor will not come through.	Moroccan Spice Blend Lebanese Spice Blend Beautiful World Blend Finnish Spice Blend Worcester Spice Blend Novgorod (Russia) Spice Blend
👌	A small amount added at the end of cooking can be a refinement to many dishes, as you would use pepper or MSG.	Yucatan Spice Blend Sicilian Spice Blend Burgundy (France) Spice Blend Beautiful World Blend Texas Spice Blend Brazilian Spice Blend
💣	Strong and spicy. If you use very much, this may be the only thing you taste.	Oaxacan Spice Blend Yucatan Spice Blend Thai Spice Blend Hungarian Spice Blend Texas BBQ Chili Powder

SOME EXAMPLES

Beautiful World Blend
• Mix a teaspoon for each 30 grams (ounce) of mayonnaise. Add a little fresh squeezed lemon juice for a killer sandwich spread.

Yucatan Spice Blend
• Add to refried beans or black beans for an amazing flavor boost.

Sicilian Spice Blend
• Make a compound butter using 2 teaspoons of the spice and a teaspoon of finely minced shallots for each 30 grams of butter.
• Add a little along with the cheese on pizza.

Worcester Spice Blend
• Season trout with this liberally, then dust with flour before frying.
• Add to beef stews, especially Irish Stew for a dramatic boost.

Novgorod (Russia) Spice Blend
• Mix 1 tablespoon per 120 grams (4 ounces) of smetana or sour cream, plus 1 teaspoon dried chives or dill as a chip dip.

HAMBURGER HELPERS

Forget that cringe-worthy commercial product. You can create instant meals using ground meats, easy-prep ingredients and these seasonings for fast meals that taste better than they have a right to.

Thai Chicken Curry - Ground chicken, canned potatoes, curry powder, Thai Spice Blend, coconut milk or dairy half & half.

Chicken Francaise - Ground chicken, paprika, Provençal Spice Blend, egg whisked with cream, top with lemon juice.

Stroganoff-esque Beef - Boil pasta. Fry beef, puréed tomato, onions, canned mushrooms, red wine, Burgundy and Novgorod Spice blends. Finish with sour cream and dill. Mix with noodles.

Goulash - Ground beef, paprika, puréed tomatoes (passata), canned carrots, Hungarian Spice Blend. Finish with sour cream.

Mexi-rolé - Ground beef or pork, Oaxacan Spice Blend, puréed tomatoes (passata), canned corn, cheddar cheese. Broil to finish.

*You can improve any of these recipes with fresh herbs on top.

AUSTRIAN SEASONING - *QUICK VERSION*

This was derived from seasonings that were used by Wolfgang Puck many years ago when I worked with him for a short time. This is the version that appeared in a couple of my videos on YouTube because it is simpler to make than the full version (see next page).

1 T	Poppy Seeds
1 1/2 teaspoons	Caraway Seeds
1 1/2 teaspoons	Lovage, dried
1 1/2 teaspoons	Sage, dried
1 1/2 teaspoons	Marjoram, dried
1 1/2 teaspoons	Green Peppercorns
1 T	Dark Brown Sugar
1 T	Salt

The poppy seeds here will help make up for some of the missing flavors in the full version, but they will also lend a slightly grainy texture to some dishes, so you have to decide whether or not to add them based on the nature of the dish. They blend in fine with meat dishes.

POPPY SEEDS AND DRUG TESTING

Although poppies are the source of the narcotic opium, the seeds used in cooking do not contain enough of the psychoactive component to be considered a drug because it would take more than you could possibly consume to have any effect. Aside from spice blends and seeds on bagels, poppy seed paste is used as a thickening agent in some Middle Eastern and Indian foods, and unfortunately poppy seeds are known to give a false positive in drug screenings. The good news is that the chemical normally clears from your system within 48 hours, so as long as your drug test is not within that time period you should be fine. Still, it is advised to allow 5 days before testing just to be safe.

AUSTRIAN SEASONING - *FULL VERSION*

If you are making the roasted duck legs, this version of the seasoning is better for both flavor and color, as it will produce a lovely brown on the skin of the duck legs that you won't get with the quick version.

50g (1.75 oz)	Green Apple, ripe
2-3 whole	Juniper Berries (optional)
2 teaspoons	Coarse Salt
2 teaspoons	Caraway Seeds
1 1/2 teaspoons	Green Peppercorns
1 1/2 teaspoons	Lovage, dried (see text below)
1 1/2 teaspoons	Sage, dried
1 1/2 teaspoons	Marjoram, dried
1 teaspoon	Dark Brown Sugar, ideally Cassonade
3-4 whole	Green Cardamom pods (see text below)

Grind the salt and juniper berries in an electric spice mill to turn it into a fine powder. Slice the apple 2mm (1/12 inch) thick on a mandoline. You do not need to peel it, but remove the seeds and stem. Toss the apple slices with the salt and then roast this in a preheated oven at 120°C (240°F) for 45 minutes with fan assist on. Now remove the "mummified" slices from the oven and let cool at room temperature for at least 30 minutes. If they are not perfectly dry and brittle, then return them to the oven until they are.

Break open the cardamom pods and remove the small round seeds inside. Add these to the rest of the seasonings and discard the husks. Alternatively, use half a teaspoon of dried cardamom powder.

Grind the dried and salted apple slices in an electric spice mill with all the rest of the ingredients. As before, you can substitute equal parts dried parsley and celery for the lovage, but the result is not quite as good.

OAXACAN SPICE BLEND

I came upon the seminal key ingredient to this when in Taxco, Mexico, many years ago. I wondered what they seemed to be burning apples over coals. They showed me the final product, which was dry, brittle and smoky. If you have a barbecue pit and hours of time on your hands, you can produce these with indirect heat. Otherwise your oven will work fine. This won't have the smokiness, so you'll have to add that in another way in the recipe, such as with liquid smoke, or by smoking the meat.

2 whole	Green Apples, ripe
2 whole	Dried Red Serrano Chilies
1 T	Cumin Seeds
2 teaspoons	Oregano, dried
2 teaspoons	Piloncillo Oscuro
	or substitute Muscovado sugar
1 1/2 teaspoons	Cocoa Powder, unsweetened
1 teaspoon	Cloves, whole (the spice - not garlic)
1 teaspoon	Orange Peel, dried (commercial product)
1/2 teaspoon	Cinnamon, ground
1 t	Coarse Salt

This is not about being a hot chili blend. This is what fills in all of those lovely background notes in many of the dishes of that region and makes an abbreviated version possible. It's not just for Mexican food, either. Try adding a teaspoon to a beef stew to really pick up the flavors!

Slice the apples into six wedges each. Place them on a wire rack in the oven, skin-side down. Roast at 120°C (240°F) for 2 hours with fan assist on. Then lower the temperature to 80°C (160°F) and continue drying for another 2 hours, still with fan assist on. Remove them and allow to stand overnight at room temperature. During this time they should become brittle so that they can be powdered. If not, dry them in the oven longer.

Combine the slightly burnt apple pieces in an electric spice mill with all of the rest of the ingredients. Grind to a powder and bottle.

YUCATAN SPICE BLEND

Because of the ground bay leaf, this is best used in cooked dishes. It really comes alive after cooking with a taste unlike any other. You can substitute 2 dried red Serrano chilies for the Chipotle chili with acceptable results, but it won't be as hot or complex in taste.

60g (2 oz)	Onion, peeled
30g (1 oz)	Corn Flour, fine (not polenta or grits)
20g (0.7 oz)	Prunes, pitted
3 T	Cilantro, fresh
1 T	Orange Zest, freshly grated
3 slices	Salted Dried Limes (page 21)
2 whole	California Laurel, dried (or Bay Leaves)
1/2 - 1 teaspoon	Dried Chipotle Chili (see note below)
1 teaspoon	Coarse Salt
3/4 teaspoon	Nutmeg, ground
1/2 teaspoon	Baking Soda (not baking powder)

The measurement of dried chipotle chili is after it has been ground to a powder in a mortar. A full teaspoon will make this very hot. Cut the onion into 3mm (1/10 inch) slices on a mandoline (or carefully with a knife). Place on a Silpat in a preheated 210°C (400°F) oven for about 20 minutes. They should be brown and just barely turning black.

Put a stainless steel pan on a medium heat (#6) and add the corn meal, salt and baking soda. While this is heating, dice the prunes finely. Toast the corn meal with frequent stirring for about 7 minutes until it just starts to darken (IR temperature 175°C/350°F). Now add the diced prunes. Lower the heat slightly (#5). Continue cooking with frequent stirring for 4 minutes. Now add the orange zest and cilantro. Continue stirring for another 5 minutes. Now crumble in the bay leaf, dried lime slices, dried chilies and browned onions. Lower the heat (#3) and cook 3 minutes. Transfer to a bowl to cool. Add the nutmeg. Put gloves on and work the mixture between your fingers well, breaking up the larger pieces. Working in batches, pass the contents through an electric spice grinder. Now rub it through a sieve. Discard what won't pass through. Transfer to a jar and use within a month.

MOROCCAN SPICE BLEND

The aroma of this mixture will instantly transport you to a Moroccan spice vender shop.

One of my older YouTube videos is *Jewel Box Chicken Tagine* (recipe notes are in Volume 1, page 222). Add 2 to 3 teaspoons of this seasoning mix to the recipe and the flavor will be much deeper and intense.

This is especially harmonious with carrots, spinach, rabbit and lamb dishes.

2 t	Lemon Peel, dried (commercial product)
2 t	Turbinado Sugar
1 1/2 t	Coriander Seeds
1 t	Anise Seeds (not star anise)
1 t	Cayenne Pepper
1 t	Coarse Salt
3/4 t	Nutmeg
3/4 t	Cinnamon, ground
3/4 t	Cumin Seeds
3/4 t	Turmeric
1/2 t	Smoked Paprika (Pimentón)
1/2 t	Black Peppercorns
1/2 t	Ginger, ground
1/2 t	Saffron threads

Simply combine all of the ingredients in an electric spice mill and grind to a powder.

If you taste this directly, you'll notice how the flavor continues to evolve in your mouth over several minutes or so, becoming quite pleasantly warming with the essence of north African flavors.

THAI SPICE BLEND

As before, some of the ingredients here are not authentic, but the resulting mixture is an incredible flavor booster to many dishes from this part of the world. Use it in the same way you would Garam Masala in Indian curries.

2 T	Corn Flour
1 T	Peanuts, shelled (raw)
2-3 whole	Dried Red Thai Chilies (no stem)
1-2 slices	Salted Dried Limes (page 21)
1 t	Curry Powder, ideally MDH brand from India
1 t	Garlic Powder
1 t	Coarse Salt
1/2 t	Thyme, dried
1/2 t	MSG
1/2 t	Galangal, ground (or substitute ground ginger)

Heat a nonstick skillet on a medium-high setting with the flour and peanuts in the pan. Stir frequently until it darkens slightly and just starts to smoke. This will take 6 to 10 minutes to get to that point. Remove from the heat and transfer to a bowl to cool for a few minutes. Now put in an electric spice mill with all of the rest of the ingredients and grind to a powder.

Try making a Thai flavored omelette by whisking together 1 teaspoon of this seasoning with 2 whole eggs, a tablespoon or so of freshly minced cilantro and some minced scallion. After it has cooked, add a little lime juice. There is no such dish in Thailand, but you'll think there is.

SICILIAN SPICE BLEND

Italian foods are rich in natural MSG, and anchovies are one of the "worst kept secrets" of Italian chefs. Dried anchovies are a very popular beer snack in Russia and readily available at every corner store. Elsewhere you may have to look online, or in a Russian specialty market.

This works well with many pasta dishes. As you would expect, this adds yet another layer of flavor to *Pasta alla Puttanesca* (Volume 2, pages 94-96), and turns the *Sicilian Stuffed Eggplant* recipe into a masterpiece (see video and notes in Volume 1, page 153). Try adding 1 to 2 teaspoons of this seasoning to the Chicken Marsala recipe (Volume 2, pages 136-137, at Step 11) to take it to the next level.

20g (0.7 oz)	Amaranth Flour
15g (0.5 oz)	Dried Anchovies
1 t	Coarse Salt
2 t	Red Chili Pepper Flakes
1 1/2 t	Orange Peel, dried (commercial product)
1 1/2 t	Basil, dried
1 t	Garlic Powder
3/4 t	Mint, dried
1/2 t	Parsley, dried
1/2 t	Cloves (the spice)
1/2 t	Black Peppercorns
1/4 t	Saffron threads

Heat a nonstick skillet on a medium-high setting. Put in the flour, dried anchovies and coarse salt. Stir until the flour darkens visibly. If you are using amaranth flour, there won't be any visible darkening for a while and then it will suddenly darken. You need to get it off the heat promptly at this point to keep it from burning. Transfer it to a bowl and allow to cool at room temperature for 10 minutes. Grind in an electric spice mill and then pass the mixture through a sieve. Discard any fragments that didn't go through the sieve. Add about 2 tablespoons of the powder back into the spice mill with the rest of the ingredients. Grind to a powder. Combine in a jar with a lid. Put the lid on and shake to mix.

WORCESTER SPICE BLEND

Reminiscent of Worcestershire sauce in the list of ingredients. The taste is much hotter than you would expect, but it is not a chili pepper type of heat. The flavor is dominated by isothiocyanates developed during the cooking process. This is the same organic functional group that gives horseradish and wasabi their hot pungent flavor. It is balanced, slightly fruity and impossible to describe.

In the restaurant where I created this, it became nicknamed "The Fishin' Magician", because it's magical on fresh pan-fried fish. Don't limit it to fish, though. It has magical powers on meat stews and potatoes, too.

45ml (1.5 oz)	Red Wine Vinegar
22ml (0.75 oz)	Distilled White Vinegar, ideally 8-12% acid
22g (0.75 oz)	Barley Flour
15g (0.5 oz)	Garlic, sliced thin
1 teaspoon	Sumac, ground
1 teaspoon	Salt
7g (0.25 oz)	Dried Anchovies
2 teaspoons	Dark Brown Sugar, ideally Muscavado
1 teaspoon	Onion Powder
1/2 teaspoon	Tarragon, dried
1/2 teaspoon	Black Peppercorns

Combine the vinegars, barley flour, garlic, sumac and salt in a nonstick pan over a medium heat. Stir until it is a sticky paste. Transfer to a ceramic baking dish and dry at 120°C (250°F) with fan assist ON for 30 minutes. Stir the contents as best as possible, then return to the oven at a lower temperature of 90°C (180°F) with fan assist ON for another hour. During this time, take it out and break up the mass between your fingers every 10-15 minutes. Nearly all of the moisture should be gone now. Grind in an electric spice mill, then return it to the same ceramic dish for a final 15 minutes in the oven at the same previous temperature. Cool to room temperature for a few minutes, then put it back in the electric spice mill along with the other ingredients. Grind to a powder and pass through a sieve. Discard solids that won't pass. Store in a glass jar in the refrigerator. Reacts with plastics and decomposes at room temperature over time.

LEBANESE SPICE BLEND ⬧

This intriguing spice mix is especially well suited for Middle Eastern eggplant and chickpea dishes. Add to Moussaka for much greater depth. This is also a great addition to the Emerati Roast Camel recipe (Volume 1, page 112).

30g (1 oz)	Gram Flour (ground dried chickpeas)
15g (1/2 oz)	Hazelnuts
2 teaspoons	Cumin Seeds, whole
1 teaspoon	Salt
1/2 teaspoon	Coriander Seeds, whole
2 whole	Dried Piquin Chilies, or 1 Dried Red Serrano
4-5 slices	Salted Dried Limes (page 21)
2 teaspoons	Brown Sugar, ideally Muscovada
1 teaspoon	Sumac
1 teaspoon	Thyme, dried
1/2 teaspoon	Garlic Powder
1/2 teaspoon	Black Peppercorns

Coarsely crush the hazelnuts in a mortar and pestle, or under the back of a heavy pan against the counter. Don't powder them - just break them down to pieces half the size of a pea. Place a stainless steel pan on a medium heat. Put in the gram flour, crushed hazelnuts, cumin seeds, salt, coriander seeds and dried chilies. Stir frequently for about 10 minutes until darkening is accelerating and a little smoke is starting to escape. Now remove the pan from the heat. Continue stirring frequently for about 3 minutes as the mixture cools slightly. Now add the sumac, garlic powder and black peppercorns. Stir for another minute or so, then transfer to a bowl to cool down close to room temperature. Crush the lime slices between your fingers and add them in along with the thyme and brown sugar. Stir to mix. Working in three batches, grind the mixture in an electric spice mill to a powder. Now put the mixture back in the stainless steel pan and heat once again, stirring almost constantly until the mixture is quite hot. Pour it off to a bowl and allow it to cool before putting it in a jar. If you are going to keep it for more than about 3 days, then store it in the refrigerator.

HUNGARIAN SPICE BLEND

Instead of adding simple paprika near the end of cooking, as is very commonly done in Hungarian home cooking, use this mixture to take a classic recipe out of the Old World of mild flavors and into the New World of robust intensity.

This goes especially well in dishes with sour cream (or smetana) as a counterpoint.

2 T	Paprika, Hungarian Hot (see page 36)
1 whole	Dried Sweet Red Chili (Volume 1, page 23)
2 teaspoons	Dried Carrots, ground
1 teaspoon	Fennel Salt (recipe on page 21)
1 teaspoon	Thyme, dried
1 teaspoon	4-Mix Peppercorns (black, white, green, pink)
1 teaspoon	Coarse Salt
1/2 teaspoon	Onion Powder
1/2 teaspoon	Ginger, ground
1/4 teaspoon	Nutmeg, ground

Simply combine everything in an electric spice mill and grind to a powder. It is best if used within a week, but still okay for up to about a month, after which the flavor has faded quite a bit.

FOOTNOTE: In Hungary, the same peppers used to make different types of paprika (see page 36) are also sold in toothpaste-like tubes as purées. Unfortunately, like many interesting Russian products, few are exported.

BURGUNDY (FRANCE) SPICE BLEND

The magic of this seasoning mix (made with black currants) appears when you add it to any red wine sauce, or dish cooked with red wine.

60g (2 oz)	Black Currants, frozen
15g (0.5 oz)	Hemp Flour (see note on page 280)
1 1/2 teaspoons	Coarse Salt
1 teaspoon	Turmeric
3/4 teaspoon	Coriander Seeds
1/4 teaspoon	Cloves (the spice)
6 whole	Black Peppercorns
1 teaspoon	Mustard, dry
1 teaspoon	Parsley, dry
1/2 teaspoon	Chives, dry
1/2 teaspoon	Garlic Powder
1/2 teaspoon	MSG
1/4 teaspoon	Cayenne Pepper

Spread the frozen black currants on a Silpat and roast in a 120°C (250°F) oven for 1 1/2 hours. Remove to cool at room temperature for at least 30 minutes, where they will continue to dry out more.

Put a stainless steel pan on a medium heat. Add the roasted black currents, hemp flour, coarse salt, turmeric and coriander seeds. Stir frequently and watch the temperature with an IR thermometer. When it reaches 60°C (150°F) begin counting 5 minutes while you continue stirring frequently. Adjust the heat to keep it around 85°C (185°F) and below 90°C (195°F) absolute maximum. When 5 minutes has passed, add the cloves and black peppercorns. Now increase the heat to medium-high (#7 out of 10) to raise the temperature to 100°C (210°F) while you continue stirring. Now remove the pan from the heat and stir for 2 more minutes. Transfer to a bowl and add the other ingredients. Stir occasionally while the mixture cools near room temperature. Now grind to a powder in an electric spice mill.

PROVENÇAL (FRANCE) SEASONING BLEND

The inspiration for this seasoning mix is verjus, also spelled verjuice. This is the juice of unripened green grapes that used to common in Europe for deglazing pans and adding sourness in a deft manner (as opposed to lemon juice or vinegar). Sorrel was often included in the Middle Ages. Verjus is still used today in some regions of France, but understand that was only the inspiration for this blend. It is not a substitute for verjus. It adds another dimension and drama to dishes that are otherwise ordinary.

90g (3 oz)	Green Grapes, seedless
60ml (2 oz)	Madeira or Alsatian Gewurztraminer
15ml (0.5 oz)	Sherry Vinegar
30g (1 oz)	Pea Flour (Peasemeal), or regular flour
2 T	Sorrel Leaves, freshly minced (optional)
3/4 teaspoon	Grains of Paradise
	or (not as good) 1/2 t Black Peppercorns
1 teaspoon	Coarse Salt
1 teaspoon	Rosemary, dried

Aside from being a natural for chicken, grouse and pheasant dishes, it is also often a sparkling touch in pasta dishes that have cream.

Cut each grape in half. Put them in a small saucepan with the wine and the sherry vinegar. Heat at a medium temperature to a slow boil. Lightly crush the grains of paradise (or black peppercorns) with a mortar and pestle and mix with the flour thoroughly. Stir in the minced sorrel, too (if you are using it). You will have sticky globular masses in the flour. This is normal. Scrape the mixture into a small baking tray. Dry in a 90°C (180°F) oven with fan assist on. Remove from the oven briefly to stir every 30 minutes. Then allow it to cool for 5 minutes or so. Crush the lumps between your fingers as much as you can. Return the pan to the oven for 30 more minutes. Now let stand at room temperature for about 15 minutes. Put into an electric spice mill along with the rosemary and coarse salt. Grind to a powder.

FINNISH SPICE BLEND ⬡

This can be used as a souring agent like amchoor is used in Indian cooking. Scandinavian food is generally quite simple, so this is far from traditional. Especially suited to salmon. More about this in Volume 4.

2 T	Oats, whole (steel cut)
30g (1 oz)	Dill, fresh
20g (0.7 oz)	Oat Flour
50g (1.75 oz)	Lingonberries, frozen
1 teaspoon	Coarse Salt
1 teaspoon	Coriander Seeds
1 teaspoon	Sage, dried
15ml (0.5 oz)	Jägermeister Liqueur

Place the whole oats (non-instant oatmeal) into the bottom of a stove top smoker to use as fuel. Put the dill into the top. Place on a burner at maximum heat with the cover cracked open until smoke begins to come out. Now reduce the heat to medium-low (#4 out of 10) and wait 5 minutes, still with the cover cracked open (not how it is usually operated). Remove the smoked dill to a bowl and cover with cling film until it is needed. Reserve the burned oats from the smoker - do not discard!

In a stainless steel saucepan on a medium heat, cook the oat flour, lingonberries and salt together, stirring frequently. Mash the berries first to a paste consistency, then gradually drying them out. After about 15 minutes when it is crumbly, add the coriander seeds and the sage. Continue cooking for another 5 minutes, stirring frequently to dry the mixture out further. Now add the smoked dill and turn the heat off. Stir for another 2-3 minutes to get rid of excess moisture. Transfer the entire mixture to a food processor and grind to a pink crumbly powder that is slightly moist. Spread this out on a ceramic baking vessel and dry in the oven at 90°C / 180°F with fan assist ON for 30 minutes. Stir and return to the oven for another 20-30 minutes. When it is almost dry, add the Jägermeister liqueur and return to the oven for about 15 minutes more, during which time the mixture should finally darken some. Now cool to room temperature and grind in an electric spice mill with a teaspoon of the charred oats from the smoker.

BEAUTIFUL WORLD BLEND

See description for *Beau Monde* on page 30. This has such universal use that I couldn't name it after a single country. Although it's especially well suited to egg dishes, don't blend it in or the eggs will turn grey. This is also a great seasoning to add to the rub for the fat cap on prime rib roasts, and all the more when served with Yorkshire Pudding.

30g (1 oz)	Amaranth Flour
15g (0.5 oz)	Arugula (also known as Rocket), wilted
2 T	Dried Porcini Mushrooms
1 teaspoon	Parsley, dried
1 teaspoon	Marjoram, dried
1 teaspoon	Celery, dried leaf or seed
1 teaspoon	Carrot, dried (optional)
1 teaspoon	MSG
1 teaspoon	Salt
1 teaspoon	Sugar
1/2 teaspoon	Garlic Powder
1/2 teaspoon	Crushed Red Pepper Flakes
1/4 teaspoon	Ginger, ground
1/4 teaspoon	Cinnamon, ground

Grind the dried porcini mushroom with a mortar and pestle. Measure out the 2 tablespoons of powder and add it to the bowl of a food processor along with the amaranth flour, arugula and salt. Process as fine as possible, repeatedly scraping up the bowl as necessary. Older wilted arugula works best for this - don't use super fresh.

Heat a large (28-30cm / 11-12 inch) nonstick skillet on a medium-low flame (#4 out of 10). Add the ground mixture and stir frequently for 15 minutes, maintaining the temperature around 88°C (190°F) by tracking with an IR thermometer. Now the arugala should be completely dry and crumble easily between your fingers, If not, then increase the temperature a little and complete the drying. Now transfer to a bowl and add all of the other ingredients while it is still warm. Stir together. When the mixture is near room temperature, grind in an electric spice mill to a powder. Pass through a sieve and discard any solids that remain.

TEXAS SPICE BLEND

Think of this as a backup chorus for chili powder and all Tex-Mex.

Dried Jalapeño chilies can be difficult to find, so you can substitute the dried green Serranos that I frequently call for in recipes.

45g (1 1/2 oz)	Shallots, peeled
30g (1 oz)	Oat Flour
2 T	Rosemary, fresh - minced
1 T	Cumin Seeds
1 1/2 teaspoons	Salt
1/2 teaspoon	Mustard Seeds, ideally black
1 whole	Dried Red Serrano Chili
1 teaspoon	Black Peppercorns
6 whole	Allspice, dried
2 teaspoons	Smoked Paprika (Pimeñton)
2 teaspoons	Dark Brown Sugar, ideally Muscovada
2 teaspoons	Dried Jalapeño Chilies, ground
1 teaspoon	Oregano, dried

Cut the shallots into 3mm (1/10 inch) slices using a mandoline. Place on a Silpat and lightly salt them. Roast in a preheated 200°C (390°F) oven for about 25 minutes. They should be starting to brown, but still fairly soft.

Put shallots in a stainless steel pan with the oat flour, rosemary, cumin seeds, salt and black mustard seeds. Heat on a medium setting until the temperature reaches 95°C (205°F) and then begin watching the time. <u>Stir frequently during this entire procedure.</u> Adjust stove heat to maintain this temperature. After 5-6 minutes increase the heat to reach 120°C (250°F). Maintain this temperature for 5-6 minutes. Now crumble in the red chili pepper. Add the black peppercorns and allspice. Increase heat to reach 150°C (300°F). Now add the smoked paprika and brown sugar, and stir constantly for about 2 minutes. Transfer to a bowl to cool. As soon as it is cool enough to handle, rub the mixture between your fingers to break up as much of the solid chunks as you can. Do this for several minutes. When it is close to room temperature add the oregano and dried jalapeño chili pepper. Working in batches, grind in an electric spice mill to a powder.

TEXAS BBQ CHILI POWDER

Not to be confused with the Chili Powder formula on page 192 of Volume 2. This is not as hot and is tailored more to BBQ recipes. Note that this is a compound spice mix and you will need the Texas Spice Blend from the previous page to make this.

In addition to barbecue sauces and dry rubs, this is the ideal seasoning for baked beans.

1 T	Paprika (see note below)
1 whole	Dried Red Serrano Chili
1/2 - 1 1/2 teaspoons	Dried Chipotle Chilies (see note below)
2 teaspoons	Dark Brown Sugar
1 1/2 teaspoons	Coarse Salt
1 1/2 teaspoons	Cumin Seeds
1 teaspoon	Oregano, dried
1 teaspoon	Thyme, dried
1 teaspoon	Black Peppercorns
1/2 teaspoon	Garlic Powder
1/2 teaspoon	Basil, dried
1 T	Texas Spice Blend (see previous page)

Instead of commercial paprika, for better results you can use the equivalent amount of ground dried sweet red chili peppers (see page 23 of Volume 1).

The dried chipotle chilies need to be crushed in a mortar first to facilitate measuring the quantity in teaspoons. They add a great deal of heat, but also flavor, smokiness and complexity. If you want it milder, then you can use less of this ingredient. Also, see the notes on page 282.

Using an electric spice mill, grind the dried chilies together with the salt and sugar. Now add the rest of the ingredients except the Texas Spice Blend, and grind to a powder. Transfer to a glass jar and add the Texas Spice Blend. Shake to combine.

BRAZILIAN SPICE BLEND

Here's the quintessential example of how these international spice blends serve to fill in background notes. The aroma of this is unlike anything you would imagine—almost magically unpredictable—yet it incorporates key flavors in cooking from this part of the world. So, while no Brazilian would use such a concoction, the results are amazing.

15g (0.5 oz)	Oat Flour
15g (0.5 oz)	Hemp Flour
30g (1 oz)	Cashew Nuts, shelled but whole
1 1/2 teaspoons	Coarse Salt
1 whole	Dried Green Serrano Chili
15g (0.5 oz)	Basil Leaves, freshly picked and minced
1 T	Rapadura Sugar, or Muscovado
1 1/2 teaspoons	Coffee, instant freeze-dried
1 teaspoon	Paprika
1/4 teaspoon	Cinnamon, ground

Place a stainless steel pan on a medium heat with the two types of flour, the nuts and the salt. Stir, toasting both the flour and the cashews simultaneously. Don't try to work too fast. Be patient. This should take 10-12 minutes. Now crumble the dried chili into the pan and continue stirring for another couple of minutes. There should be a sharp aroma from the chili as it browns slightly, but if smoke is billowing out then you burned it (start over again). Now remove the pan from the heat and stir in the basil leaves and the paprika. Let the residual heat in the flour and other ingredients dry out the basil. Within 3-4 minutes the basil should be dry and crumble to the touch. Now remove the contents of the pan to a bowl and allow it to cool for 10-15 minutes. When it is no longer hot (warm is okay), stir in the sugar, coffee and cinnamon. Working in batches, grind this in an electric spice mill to a powder. Store in the refrigerator.

NOVGOROD (RUSSIA) SPICE BLEND ⬥

This one is almost in a category of its own. It was created as a way to bridge western flavors to Russian tastes. Incorporated just below the level of conscious taste perception, this introduces a subconsciously familiar taste to Russians that improves scores on taste tests. If you are not Russian, you may find it strange. This works well in small amounts in many Russian recipes, as you might add pepper to a dish.

15g (1/2 oz)	Buckwheat Flour (don't substitute)
1 teaspoon	Potato Starch
1 teaspoon	Coriander Seeds
1 teaspoon	Black Peppercorns
2 teaspoons	Yeast, dried
1 teaspoon	Salt
1 1/4 teaspoons	Khmeli-Suneli (Georgian spice mix)
1/2 teaspoon	Dill, dried
1/2 teaspoon	Garlic Powder
1/4 teaspoon	Parsley, dried

In a dry nonstick skillet on a medium heat (#5 out of 10), toast the buckwheat flour, potato starch, coriander seeds and black peppercorns for about 6 minutes, stirring frequently. If you have an IR thermometer, wait until the temperature is 150°C (300°F). Now add the yeast and the salt. Continue toasting and stirring for another 4-5 minutes, until the temperature reaches 175°C (350°F). Scrape out the contents to a bowl and mix the khmeli-suneli, dried dill, garlic powder and dried parsley in. Let it come to room temperature before grinding in an electric spice mill to a powder.

As a side note, this forms an Anti-Resonant flavor pairing (Volume 2, page 26) with roasted green bell peppers. The combination is something quite unexpected.

✦

OTTOMAN SPICE BLEND

For hundreds of years the Ottoman Empire was trading spices and food products across half the globe. Turkish cuisine is descended from this, but it is distinctly different, too. This spice blend is based on seasonings I found in ancient recipes. Although their proportions were not stated directly, this blend is well balanced with a very long finish. Don't judge it by how it tastes directly. It needs to be a flavoring in a dish.

1 T + 1/2 teaspoon	Fennel Salt (see note below)
1 1/2 teaspoons	Black Peppercorns, whole
1 1/2 teaspoons	Oregano, dried
1 teaspoon	Paprika, ideally Csemege (page 36)
1 teaspoon	Fenugreek Seeds, whole
1/2 teaspoon	Coriander Seeds, whole
1/2 teaspoon	Cumin Seeds, whole
1/2 teaspoon	Mustard Seeds, whole
1/2 teaspoon	Sage, dried
6 whole	Cloves (the spice)
1/8 teaspoon	Cayenne, dried powder

The directions for making Fennel Salt are on page 21. Alternatively you can substitute 2 teaspoons of regular salt, but the flavor will not be nearly as interesting since this is the principle ingredient.

In a cast iron or nonstick skillet, lightly toast the whole fenugreek. Be careful not to burn it, as it burns very easily. As soon as it darkens slightly, it is done. Pour it out and add the coriander seeds, cumin seeds and mustard seeds to the same pan. About ten seconds after the mustard seeds begin to pop, pour out the contents of the pan to a bowl and let the spices cool to room temperature. Now grind all of the ingredients together in an electric spice mill. Store in a sealed jar away from sunlight.

Ottoman Crumble

Mix a tablespoon of the Ottoman Spice Blend (above) with about 30 grams (1 oz) of shelled pistachios. Carefully and briefly toast on a hot pan. Crush with a mortar and pestle. Don't turn it into a fine powder, but make sure there are no large pieces. See the next page about applications.

✦

COATINGS & CRUMBLES

What distinguishes a spice blend from a coating or a crumble? A spice blend is intended to be cooked in with foods. A coating must not contain anything that will burn easily. For example, the *Louisiana Crumble* (page 245) can't work as a coating because it will burn long before the rest of the food would brown. A crumble is something that can be used with little or no cooking, such as the topping on an *Au Gratin* dish that will only be exposed to the broiler for a brief time.

Crumbles are one of those "star chef" secrets that rarely get mentioned explicitly. They can produce a world of flavor that will amaze even the most jaded foodie. The Ottoman Crumble on the previous page is a perfect example. Try a sprinkle on whatever you cook for a while to get a feel for how it plays. You will discover some incredibly delicious and unexpected combinations to impress your guests. For example, *Penne in Arrabbiata Sauce* (page 210)... and before you scream, remember Marilyn's quote on page 216.

SYMBOL	MEANING	APPLIES TO...
⦿	Also recommended as a seasoning to be mixed in with ground meat for meatballs, meatloaf, and poultry stuffings.	Fuse Coating Lovecraft Coating South India Coating Sunflower Seed Crumble
(salt shaker)	This coating does not provide enough salt. You must add salt to the final dish (to taste).	RBP Coating Lovecraft Coating Louisiana Crumble South India Coating
❄	Needs to be stored in the refrigerator. This can be frozen for prolonged storage, but the taste will be less pronounced.	Fuse Coating Louisiana Crumble

RBP COATING

This was developed for fish originally, but if you make it with hard red winter wheat instead of semolina, you will get a very interesting product that gives the flavor of legs and thighs (dark meat) chicken to breast portions. If you love the flavor of dark meat but you want the health benefits of boneless, skinless chicken breast, then is something you need to try out. Otherwise (with semolina) this is a straightforward way to boost the crust on steaks and roasts—just sprinkle a little on (don't dredge the meat in it).

Unlike the other coatings here, this one is primarily about texture and more subtle in the flavor.

45g (1.5 oz)	Semolina
	or Hard Red Winter Wheat (see note above)
15g (0.5 oz)	Red Bell Pepper, finely diced
1/2 teaspoon	Baking Soda (not baking powder)
1/2 teaspoon	Citric Acid
1/2 teaspoon	Salt

Begin by heating a stainless steel sauce pan (do not use nonstick or cast iron for this) on a medium heat (#6 out of 10). Add the hard red winter wheat and the baking soda. Stir frequently for the next 5-6 minutes. There will be a fairly rapid increase in brown coloration around this time. Now add the diced bell pepper and turn off the heat. You will need to use a metal spatula to scrape the bottom and keep the mixture moving so as to dry out the bell pepper. After another 5 minutes add the citric acid and the salt. Continue stirring frequently for another 5 minutes until the mixture has cooled down. Now put the mixture into an electric spice mill, working in two batches. Pour the contents back into the same pan (wipe it out first) and heat once again on medium (#5 out of 10). Stir frequently for another 5 minutes or so until you begin to see darkened flecks. Now pour it out into a bowl and allow it to cool to room temperature. It should be bone dry now. Store it in a sealed jar.

FUSE COATING ⊙ ❄

One of the giants of the food service industry is Hormel. The company recently introduced their "Fuse" pre-cooked meat patties that make use of this approach, incorporating dried vegetable and fruits as seasonings. This recipe is based on that formula without any chemical additives.

Useful for many meats and even vegetables. When it comes to cheeseburgers, this is a fantastic product if you are using Swiss cheese or Muenster. If you are using cheddar, then I suggest the Lovecraft coating on the next page instead.

120g (4.2 oz)	Red Cargo Rice (see below)
60g (2 oz)	Onions
50g (1.75 oz)	Spinach, fresh
40g (1.4 oz)	Dried Cherries or Cranberries
30g (1 oz)	Celery
2 teaspoons	Salt
30g (1 oz)	Bread Crumbs
1/2 teaspoon	Black Pepper, finely ground

Boil the raw rice in an excess of salted water for 5 minutes, then cover and maintain at a very low heat for another 20 minutes. Drain on a sieve. The rice should be firm and just a little too al dente to eat. Let it cool to room temperature. Now weigh out the 120 grams (4.2 ounces). In a food processor, combine the rice, onions, spinach, dried fruit, celery and salt. Grind as fine as possible. Pour the mixture out onto a large nonstick pan on a medium heat (#5 out of 10). Stir while heating to evaporate off as much of the moisture as you can. After about 20 minutes, it should be crumbly. Now add the bread crumbs and increase the heat to medium-high (#7 out of 10). Cook with frequent stirring for 5-6 minutes, watching for the bread crumbs to turn caramel brown. Do not burn the mixture! Now transfer the contents of the pan to a bowl and remove the pan from the heat. After a couple of minutes, put the contents back into the still-warm pan and allow the residual heat to dry the mixture further, stirring occasionally over the next 15-20 minutes. Now add the black pepper and grind in an electric spice mill, working in batches.

LOVECRAFT COATING

Another clone of a commercial product not sold directly to consumers. This one is based on a British steak sauce. It is well suited to lamb, but the real magic is with ground beef. In the same way that the *Beautiful World* seasoning (page 233) takes a burger in the direction of fast food, the Lovecraft Coating adds the flavors of a fine steakhouse burger, especially if you can cook it over glowing coals. A prime example of tertiary magic.

60g (2 oz)	Semolina
60g (2 oz)	Dates, pitted and diced
30g (1 oz)	Bread Crumbs
15g (0.5 oz)	Garlic Cloves, chopped
2 whole	Turkish Bay Leaves, crumbled
1 teaspoon	Coarse Salt
1/2 teaspoon	Black Peppercorns
1/2 teaspoon	Cloves (the spice)
2 teaspoons	Orange Peel, freshly grated
1 teaspoon	Onion Powder
1/2 teaspoon	Ginger, ground
1/2 teaspoon	Parsley, dried

Use a food processor to grind together the semolina, dates, bread crumbs, garlic, crumbled dry bay leaves, salt, peppercorns and cloves. Make into coarse meal. Some of the peppercorns and cloves will remain whole, which is perfectly fine. Leave them like this. Now scrape the mixture into a large nonstick skillet on a medium heat (#5 out of 10). Stir frequently and, if possible, monitor with an IR thermometer. The temperature will stay around 100°C (210°F) for quite a while, but after about 12 minutes, you will see the temperature go up to about 120°C (250°F). After 1-2 minutes at this temperature, add the grated orange zest. Continue stirring frequently. Maintain at this temperature for 3 more minutes, then turn off the heat from the stove. Continue stirring as the temperature gradually goes down. It should be brown at this point, but not burnt in any way. After 5-6 minutes, stir in the onion powder, dried ginger and parsley. Allow it to continue cooling on the warm pan for another 10 minutes. Now grind it in an electric spice mill, working in batches.

SOUTH INDIA COATING

This is a rare instance of when I call for curry leaves. Applications for this as a coating will be introduced in Volume 4. You can also use this in place of Amchoor (see Volume 2, page 35) to bring more depth of flavor. It may also be added to any biryani dish with excellent results, and a small portion benefits many curries and Indian vegetarian dishes. Put it in at the same time you would add Garam Masala (recipe in Volume 1, page 134).

30g (1 oz)	Gram Flour (also known as Chickpea Flour)
1 teaspoon	Citric Acid
1 teaspoon	Coarse Salt
1 medium	Tomato
6-7 whole	Green Cardamom pods
2 teaspoons	Cumin Seeds, whole
4-5 T	Curry Leaves, dried and crumbled well

Heat a stainless steel pan (do not use nonstick or cast iron for this) on a medium heat (#6 out of 10). Put in the gram flour, citric acid and salt. Stir occasionally. Cut the tomato in halves and remove the stem and seed pockets. Squeeze it out over the sink to remove the gelatinous membrane. You want about 60 grams (2 ounces) of the "meat" of the tomato. Dice it finely. Monitor with an IR thermometer, if possible. When the temperature of the gram flour reaches about 145°C (290°F), add the diced tomato along with the cardamom pods and cumin. Stir well with a metal spatula, scraping the bottom to keep the mixture from sticking and burning. Continue cooking for 10 minutes. Now remove the pan from the heat and add the curry leaves, stirring constantly for the next 3 minutes. Allow the mixture to cool down close to room temperature. Now grind this in an electric spice mill, dividing it into two batches. Pass the ground mixture through a fine mesh sieve. A considerable portion will not pass through, but that's to be expected because of the tomato solids, cardamom husks and other plant matter. Collect the fine powder in a jar and store in a cool place. It doesn't need to be refrigerated, but best used within 2 months. You should have about 60 grams (2 ounces) in all.

ONION RING COATING

This one is really in a category by itself. To alter the texture of the onion rings, if desired, substitute additional regular all-purpose flour for the hard red winter wheat. To make it even softer, substitute soy flour for the hard red winter wheat. This can be used with other vegetables, too. Zucchini works well, but the oil will become wet much faster than with onion slices. Once the oil is wet, the cooked pieces will start to be soggy and oily. Other applications will be shown in Volume 4.

1 whole	Egg
1 T	Instant Mashed Potatoes, commercial product
1 T	Hard Red Winter Wheat Flour
1 T	Flour (see note above)
1/2 teaspoon	Salt
1/2 teaspoon	Baking Powder (not baking soda)

Whisk all ingredients together. Allow to stand for 15 minutes before use. Batter should be used within an hour of making it.

Making Onion Rings

Peel onions and slice fairly thick. Separate into rings. You will need one recipe of the batter (above) for every 1-2 onions. Only use the larger outer rings. Mix with batter while the oil heats to about 170°C (340°F). Unless you have a large commercial deep fryer, cook only a few at a time to make sure that the temperature of the oil doesn't drop too much. Fry each ring for a total of about 3 minutes. They will float to the top after a few seconds. Turn them over in the oil after half the cooking time has elapsed. Drain on paper towel and sprinkle with a little more salt.

LOUISIANA CRUMBLE

Sprinkle on top of *Au Gratin* dishes, especially. Well suited to fish and seafood. Packs loads of flavor!

60g (2 oz)	Butter
20g (0.7 oz)	Amaranth Flour
40g (1.4 oz)	Hard Red Winter Wheat, or Semolina
90g (3 oz)	Onion, chopped fine
60g (2 oz)	Celery, chopped fine
30g (1 oz)	Red Bell Pepper, chopped fine
180ml (6.3 oz)	Guinness Ale
200g (7 oz)	Tomato Purée (passata)
2 whole	Bay Leaves
1 whole	Dried Red Chili Pepper (see note below)
2 teaspoons	Thyme, dried
1 teaspoon	White Peppercorns
2 teaspoons	Coarse Salt
90g (3 oz)	Bread Crumbs (in all)
2 teaspoons	Lemon Zest
2 cloves	Garlic, chopped
1 teaspoon	MSG (or another 1/2 teaspoon salt)

Ideally use a red Serrano chili you dried yourself. Otherwise, crushed red pepper flakes. Melt the butter on a medium heat. Cook the Amaranth for 2 minutes before adding the wheat flour. Cook to make a roux, stirring frequently. This will darken faster than a white flour roux. When it is brown (about 5 minutes) add the onion, celery and bell pepper. Keep stirring frequently as the mixture dries out. When it is crumbly, add the Guinness. Continue cooking and stirring for about 10 more minutes to get it nearly dry again. Grind the bay leaf, chili pepper, thyme, white peppercorns and coarse salt in an electric spice mill. When the mixture in the pan is nearly dry, add the spice mixture. Cook for 2 minutes and then add the tomato purée. Maintain the heat on medium with frequent stirring for about 20 minutes as the mixture thickens into a semi-solid mass. Now add the lemon zest and garlic. Cook another 5-10 minutes, then stir in 60 grams (2 ounces) of the bread crumbs. After a few minutes it should be quite dry. Cool, then grind in food processor with MSG and the rest of the bread crumbs. Refrigerate.

SUNFLOWER SEED CRUMBLE

Unlike most savory crumbles, this is <u>not</u> suggested for typical *Au Gratin* dishes. It is ideal for salads and many baked vegetable dishes, though. More applications for this will be detailed in Volume 4 of this series.

90g (3 oz)	Sunflower Seeds, shelled
30g (1 oz)	Bread Crumbs
15g (0.5 oz)	Scallions, chopped
10g (0.35 oz)	Dill, chopped
2 teaspoons	Coarse Salt
1 teaspoon	Garlic Powder
3/4 teaspoon	Sugar, preferably Turbinado
1/4 teaspoon	Cayenne

In a food processor, combine the sunflower seeds, bread crumbs, scallions, dill, salt and garlic powder. Process until fine. Pour this into a large nonstick skillet on a medium heat (#5 out of 10). Stir frequently for about 10 minutes. If you have an IR thermometer, you'll notice that the bread crumbs begin to brown as the temperature finally rises above 130°C (255°F). At this point, remove the pan from the heat, but continue stirring frequently. Add the sugar and the cayenne. After the temperature is back below 70°C (150°F), leave the mixture in the warm pan to cool down to room temperature without further stirring. Transfer to a jar and store.

Killer Cocktails 3 !

As you probably know from my videos on YouTube, I am the author of *Cocktails of the South Pacific and Beyond*. The cocktail recipes presented here were all developed after the publication of that book, so there are no duplicates.

Y

LYCHEE MARTINI Nº. 2

This is an entirely different approach to the Lychee Martini explained in my cocktail book. This is especially interesting because it contains <u>no</u> actual lychee (other than the optional garnish), yet it tastes like quality lychee liqueur.

Be sure to use pure quince juice, which is available in bottles.

Combine:

60ml (2 oz)	Gin, Beefeater
45ml (3/4 oz)	Quince Juice (see note above)
15ml (1/2 oz)	Mango Syrup, Monin
10ml (1/3 oz)	Lime Juice, fresh

Shake with ice and strain into a Martini glass. Add a canned lychee on a skewer, if possible.

For a stronger tasting alternative version of this, add 15ml (half an ounce) of white rum to this recipe. Although the alcohol concentration is only slightly greater, it has a profound effect on the taste.

Y

24 KARAT GOLD CADILLAC MARGARITA

The Cadillac Margarita is a Tex-Mex invention that has a strong orange citrus taste. Most of the time all it means is you get an extra shot of Grand Marnier (often served on the side) along with a regular Margarita. That's not how to do things properly, though. This is how to do things properly:

Note that you must have Monin's *Glasco Citron* syrup, or make up your own version with lemon juice, lemon oil, lime juice and sugar. Also note that the combination of tequilas will produce a vibrant, yet smooth taste that is superior to either one on its own.

Into the shaker put:

2 strips	Orange Peel, fresh (use a vegetable peeler)
3/4 t	Sugar, white
1-2 teaspoons	Orange Juice, freshly squeezed

Muddle the orange peel well, using the granular sugar to help pulverize it. Now add:

22ml (3/4 oz)	Tequila, Jose Cuervo *Añejo* (Gold)
22ml (3/4 oz)	Tequila, Cazadores *Reposada*
15ml (1/2 oz)	Cointreau
15ml (1/2 oz)	Grand Marnier
15ml (1/2 oz)	*Glasco Citron* Syrup, Monin (see note above)
45ml (1 1/2 oz)	Lime Juice, fresh

Select either a very large margarita glass or a highball glass if you plan to serve it on the rocks. Otherwise a regular margarita glass, or even a martini glass will work. Rub the rim of the glass with one of the limes you squeezed the juice from, then rim with salt (preferably a premium salt such as Hawaiian pink).

Shake well with ice. Either pour into the glass along with the ice cubes, or strain to serve straight up. Now measure out the final dose of orange liqueur...

30ml (1 oz)	Grand Marnier

Gently pour over the back of a spoon to float it on top. Add a wedge of lime, and if serving on the rocks, then also a straw.

Ŧ

FLYING SCORPION

In case you are wondering, this is an actual species of insect that looks like a scorpion with wings. The cocktail is far more appetizing, I assure you.

If you have read my book *Cocktails of the South Pacific*, you will recall the chapter on the classic Scorpion and its various versions. Here is a high octane version that is very easy to make. Grand Marnier pulls double duty as both cognac and the orange liqueur.

Combine:

30ml (1 oz)	Gin, Beefeater
30ml (1 oz)	Grand Marnier
15ml (1/2 oz)	Light Rum, ideally Havana Club *3-Year*
20ml (2/3 oz)	Lime Juice, fresh
1 teaspoon	Orgeat Syrup

Shake with ice and pour out into a stemmed glass along with the ice cubes. If a weaker drink is desired, then add a splash of soda water. Now float...

1 teaspoon	151-proof Rum, such as Bacardi

Garnish with a tropical flower that has a few drops of almond extract on it.

OLD SCHOOL MARASCHINO CHERRIES

Back in the day when bartenders did more than just sling drinks, maraschino cherries in top of the line clubs were handmade affairs. Ideal for traditional cocktails like the Manhattan.

Combine in a small saucepan:

60ml (2 oz)	Maraschino Liqueur
2 teaspoons	Sugar
6 whole	Sour Cherries, fresh (pitted)

Bring to a boil, then pour off into a small bottle. Cover loosely until it cools back down near to room temperature. Now close the bottle and refrigerate for at least 24 hours, and ideally a week.

PEAR AND VANILLA INFUSED BRANDY

You can serve this on the rocks by itself, but it is a great addition to cocktails, particularly champagne cocktails. It is also called for in the *Speedball* recipe below on this page. You can scale this up just by multiplying all of the ingredients.

Cut a Williams or Bartlett pear into 1.5 centimeter (1/2 inch) dice including the skin, but not the stem. Put into a glass jar and add:

180ml (6 oz)	Brandy or (better) Cognac
30g (1 oz)	Sugar
1/2 bean	Vanilla Bean, split

Screw the lid on and shake to combine. Put it in a cool, dark place for between 2 and 5 days, shaking it one or two times a day.

Now press as much as you can through a fine mesh sieve. Then pass that liquid through a coffee filter to remove any cloudy bits. Store it in the refrigerator if you are not going to use it within a few days.

XXX MOONSHINE

For those craving the taste of high quality moonshine. This delivers a real kick with a surprisingly authentic flavor. Incidentally, never buy Mezcal with a worm in the bottle. That's primarily made for Americans from the worst quality ingredients. There are some great artisanal Mezcals these days if you shop around. You can also use tequila in place of this with good results.

Combine in a small jar:

180ml (6.3 oz)	Vodka
30ml (1 oz)	Pear and Vanilla Infused Brandy (see above)
30ml (1 oz)	Mezcal
22ml (3/4 oz)	151 proof Rum, such as Bacardi
1 teaspoon	Activated Charcoal, coarse (not fine powder)

Shake it up and store in a cool dark place for 24 hours. Pass through a sieve, and then through a coffee filter. Serve in a mason jar. Room temperature is traditional, but cold tastes better.

SPEEDBALL

In drug parlance, a speedball is a mixture of cocaine and heroin. It's what killed John Belushi and is also the topic of the 1970's Traffic song, *The Low Spark of High Heeled Boys*. It is highly addictive and surprisingly energizing. Your only problem will be in getting the *Adrenaline Rush*, which has been removed from the shelves of almost every store in the world except in Russia. There's nothing else you can substitute, either—I tried 22 other brands. All failed.

Combine:

60ml (2 oz)	Pear Brandy (see recipe above)
15ml (1/2 oz)	Vodka
15ml (1/2 oz)	Lemon Juice
90ml (3 oz)	Adrenaline Rush (an energy drink)
* 4 thin strips	Orange Zest

Put the pear brandy, vodka and lemon juice in a shaker. Shake with ice. Now add the Adrenaline Rush, swirl and decant with some of the ice into a chilled highball glass. Add the zest strips and a straw.

Most zester bar tools produce four strips at a time.

SARDINI

Inspired by the animated television series *Futurama* in which Zoidberg orders a martini-like drink of this name in a casino on Mars. This is a novelty cocktail, but still quite drinkable if done right.

Combine in a shaker:

140ml (5 oz)	Gin
1/2 teaspoon	Sicilian Spice Blend (page 226)

Stir. Let settle, then decant the green-tinged gin from the grainy solids. You can pass it through a coffee filter, too.

Now mix a Martini in the usual way with this flavored gin and dry Vermouth. Garnish with a sardine-stuffed olive (or two). Don't bother trying to buy sardine-stuffed olives. You'll have to stuff your own, and be sure to rinse the sardine pieces first.

DOVER DEMON

The glowing red eyes make for a fun presentation in a dark room.

60ml (2 oz)	Rum, gold (*e.g.* Flor de Caña)
2 teaspoons	Rum, 151 proof
1/2 teaspoon	Maraschino Liqueur
15ml (1/2 oz)	Lemon Juice, fresh
15ml (1/2 oz)	Orange Juice, fresh
15ml (1/2 oz)	Mango Syrup, Monin (or homemade)
1/4	Lime, fresh
2 whole	Cloves (the spice)

Stud the lime with the cloves like eyes. Put it into the metal cap of a shaker and drizzle the 151-proof rum over it. Use a culinary torch to burn the protruding tops of the cloves well. This will ignite the rum. If you dim the lights, the cloves will look like flashing red eyes. Let it burn for about 20 seconds before extinguishing it. Caution: The cap is now <u>very</u> hot. Hold in a towel to dump the contents into a shaker with all of the other ingredients. Shake with ice. Strain into a martini glass and garnish with pineapple, if desired. Bizarre but good flavor.

PALAZZO COCKTAIL

Think of this as a more accessible Negroni. Although it is probably new to you, sea buckthorn is a great counterpoint to Aperol.

Combine:

50ml (1 3/4 oz)	Gin, Beefeater or Tower Hill (an Italian brand)
15ml (1/2 oz)	Aperol (an Italian bitter orange liqueur)
2 teaspoons	Sea Buckthorn Syrup
1 teaspoon	Sweet Italian Vermouth, Martini & Rossi

Combine all ingredients in a shaker. Shake with ice. Into a chilled lowball glass with 3-4 ice cubes, add in this order:

60ml (2 oz)	Schweppes Indian Tonic Water
1/4 teaspoon	Fernet Branca (an Italian aromatic bitter)

Strain over the contents of the shaker. Sink a lemon wedge into the drink and then add a straw.

TWO INSPIRED BY LINDT CHOCOLATES

OPEN SESAME

A decorative specialty glass adds mystery to this oddball cocktail.

Combine in a shaker:

75ml (2 1/2 oz)	Brandy
15ml (1/2 oz)	Jägermeister
1 whole	Egg Yolk
1 teaspoon	Lime Juice
1/2 teaspoon	Orgeat Syrup
1/2 teaspoon	Sesame Oil (toasted)

Shake with ice. Rub the rim of a glass with lime juice and then roll in some <u>sesame seeds</u>. Don't crust the entire rim with sesame seeds—just a few. Strain into the glass. Peel curls over the top of:

1/4 square	Lindt *Roasted Sesame* chocolate

PAY DIRT

It looks like dirt, but tastes like a million bucks! The Nicaraguan rum called for has a nutmeg-like taste that works magic. You *can* skip the lime bitters and still have a great cocktail, though.

Combine:

60ml (2 oz)	Rum, Flor de Caña *7 Year Gran Reserva*
15ml (1/2 oz)	Apricot Brandy, Bols or Marie Brizzard
15ml (1/2 oz)	Cointreau
10ml (2 teaspoons)	Lime Juice, fresh
3-4 drops	Lime Bitters, Fee Brothers (optional)

Combine all ingredients in a shaker. Shake with ice and strain into a Martini glass. Grate over the top:

1/2 square	Lindt *Intense Lime* chocolate (kept in freezer)

Grate enough to cover the entire top of the cocktail as the "dirt", then drop the rest of the square into the cocktail as the "payoff". A little edible gold leaf on top adds some great dramatic flair.

CORIANDER NASTOYKA

Nastoyka is Russian for a vodka that has been quickly flavored, usually with some kind of plant matter. Typically the vodka is mixed with the flavoring agent for 48 hours or less. I've substituted half of the honey used in the classic recipe for maraschino cherry syrup because it adds another dimension to the flavor.

Combine:

200ml (7 oz)	Vodka
1 1/2 teaspoons	Coriander Seeds
1 teaspoon	Honey
1 teaspoon	Maraschino Cherry Syrup (see note below)

Ideally you want to use the liquid from the *Old School Maraschino Cherries* (page 249), but you can substitute another teaspoon of honey, or the syrup from commercial maraschino cherries. Combine ingredients in a jar with a screw lid. Shake well. Let stand for 24 hours before passing through a sieve to remove the coriander seeds. It won't spoil, but the flavor fades over several weeks.

SWEET ITALIAN ROSE

The secret to the balance of this drink is the coriander spice that no one ever suspects.

Combine:

60ml (3 oz)	Vodka
30ml (1 oz)	Coriander Nastoyka (see above)
15ml (1/2 oz)	Sweet Italian Vermouth
15ml (1/2 oz)	Rose Syrup, Monin
15ml (1/2 oz)	Lemon Juice, fresh
1 teaspoon	Campari

Combine all ingredients except the Campari in a shaker. Shake with ice. Pour the Campari into a chilled martini glass and swirl to coat. Let the excess run out. Now strain the drink from the cocktail shaker into the glass. Garnish with one or two rose petals.

OMAKASE

The name of this drink is the Japanese term used in sushi bars meaning that it is up to the chef as to what to make. This is also a request I frequently get when tending bar, and this is often the drink that I serve to customers whom I don't know very well, because it contains elements that please almost everyone and the flavor is unexpected and bewitching. Plus there are some mild theatrics in the production of the drink. I have been asked for this recipe by a great many customers, but this is the first time I have provided it.

You can substitute mango syrup for the passion fruit syrup if you need to. It will not be as good, but at least you can get a good idea about the taste if you don't have passion fruit syrup on hand.

In the same way, the coriander bitters are expensive and only available online, so I suggest that you first try the cocktail with Angostura bitters instead. If you love it, then get the coriander bitters and see how much better it can be.

Combine:

45ml (1 1/2 oz)	Coriander Nastoyka (see previous page)
15ml (1/2 oz)	Jägermeister
15ml (1/2 oz)	Benedictine Liqueur
30ml (1 oz)	Orange Juice, fresh
1/2 teaspoon	Passion Fruit Syrup (see note above)
4-5 drops	Coriander Bitters, Bob's (see note above)

Thread an *Old School Maraschino Cherry* (page 249) onto a skewer or swizzle stick. Lightly rub the cherry around the rim of a Coupe glass, letting any liquid from the cherry drip down into the glass. Don't press so hard that you damage the cherry. Now lay the swizzle stick across the top of the glass with the cherry centered on it. Shake the ingredients with ice and strain into the glass over the top of the cherry.

NOTES

Beer & Wine Tasting Lingo

ABV An acronym for Alcohol By Volume, which is the percentage of alcohol measured in any beer, wine or other liquor. Beer averages around 5% alcohol (ABV). See Proof.

Accessible Easy to like the first time you try it.

Acetaldehyde A green apple aroma or taste. A yeast or bacterial by-product in ales sometimes.

Acetic Aroma or flavor of vinegar formed by aerobic bacteria producing acetic acid. Acetic means bacterial spoilage has occurred, except in the case of Lambic ales where this taste is expected, Not to be confused with the slightly acidic nature (tartness) expected in many types of wine.

Acidic Taken to mean excessive acidity. This can be lemon or lime in nature, which might be intentional, or the result of improper fermentation or spoilage. See Acetic and Tart.

Aftertaste The taste left on the palate after it has been swallowed. See also Finish.

Aggressive A beer with pronounced or over the top flavors. The opposite of a beer described as "smooth" or "soft"

Alcoholic A noticeable presence of too much alcohol. Often noted as hot or spicy in beers. Rarely a complaint in wines.

Almondy Aroma of marzipan. Nearly always a character in Belgian Kriek ales. Also common in Gewürztraminer and some other white wines.

Angular The taste seems to hit a few parts of your mouth like sharp corners. Especially used for high acidity wines.

AoC The highest ranking of wine in the old French system.

AoP *(Appellation d'Origine Protégée)* The highest quality wine in the new French classification system. See AoC.

Astringent Very dry aftertaste. Sometimes with a harsh grain flavor. This is common in Japanese beers.

Aroma The smell, either good or bad. See Bouquet.

Autolytic or Autolysed Aroma of yeasty or acacia-like floweriness associated with beers that have been aged too long. It can also sometimes be described as Meaty.

Balanced A beer or wine in which no one single flavor component stands out. In beers this usually refers to a hop vs. malt balance. In wine there are many components.

Banana Aromas like banana or notes of bubble gum from Isoamyl acetate; a fermentation by-product in beer. Also a note in some white wines, which is usually pleasant.

Band-aid An unpleasant medicinal aroma or taste.

Barnyard Earthy and vegetal undertones (often with hints of sweat or urine mixed in). These aromas are most common in Lambic ales. At low levels some beer devotees find this appealing. At high levels it is foul tasting. See Farmyard.

Big Intense flavor, or (mostly in wine) high in alcohol.

Biodynamic Unlike "Organic", there is no legal requirement or inspection for this designation, but the vineyard is declaring that certain guidelines were followed. There are some "new age" winemakers who use this to mean that the grapes were planted and harvested according to astrological signs and the cycles of the moon, which has no scientific basis.

Biscuity A beer descriptor often associated with strong malt. It is a slightly yeasty bread dough aroma and flavor.

Bite In beer, the initial taste of hops, tannins, high carbonation and/or acidity. This can be a positive or negative, depending on if the beer is balanced. Bite is always negative for wines.

Black Currant An aroma or flavor of black currant, raisins or cassis. Common in Bock beers and many red wines.

Body The sense of feeling in the mouth. The sense of fullness from malt or alcohol in the beer. Some common descriptors are thin, winey, delicate, light, medium, balanced, robust, full, heavy, dense, viscous, overwhelming

Boozy An exaggerated alcohol aroma. Often associated with strong ales, but also very much a matter of opinion,

Bottle Conditioned A beer in which fermentation was completed in the bottle. It contains live yeast (unfiltered) and enough sugar for fermentation to continue after bottling.

Bouquet The pleasant aromas from either a beer or wine.

Brett An abbreviated term for *Brettanomyces* yeast, which plays a role in both ale and wine production. Generally considered harmful rot in wine production, but a tiny bit is sometimes regarded as a blessing. Some describe the taste as that of cough syrup. An excess of this in beer will give the characteristic Brett taste. See Metallic.

Bright When describing the visual appearance of the beer, it refers to high clarity, very low levels of suspended solids. Lacking haze. For wine, it means clear and not opaque.

Brussels Lace When the tracks of liquid that cling to the sides of a glass after the contents have been swirled display a delicate pattern that resembles lace. See Lace and Legs.

Brut Champagne that is very dry. No sugar can be tasted.

Burnt Having an aroma or flavor of smoke or burnt wood. In beer this is caused by excessive heat during boiling (heating with a gas ring or electrical elements) or from unclean heating surfaces. Almost never applied to wines.

Buttery In wine this is usually derived from oak barrel aging, particularly for Chardonnay. In beer, a noticeable but acceptable level of Diacetyl, giving a rich, creamy mouthfeel and flavor reminiscent of butter. Too much of this flavor is considered a defect. See Diacetyl.

Butyric Aromas of rancid butter. Always a flaw in beer or wine.

Caramel Aromas or flavors of caramel, brown sugar, toffee. Most often applied to dark beers, especially Dopplebocks.

Carbonated or Carbonation The amount of carbon dioxide dissolved in the beer (usually between 4.5 and 6 grams per liter). This is what gives beer its effervescence. Some common descriptors are spritzy, sprightly, zesty, prickly, gassy, sharp, round, smooth, creamy, delicate, piquant, and even champagne-like.

Cardboard Having an aroma of wet cardboard. See Oxidized.

Cassis Usually associated with the seedy quality more than the sweetness of this fruit when used in the descriptions of wines, especially Italian Ripasso. For Belgian ale, it's fruit.

Catty Having the aroma of cat urine. See Skunky.

Chalky having a mouthfeel that is dusty, chalky, or feels like microscopic particles are in it.

Chewy For beer, the malt is overwhelming. For wine, the tannin is too drying. See Astringent.

Cheesy The aroma of cheese. A serious flaw in beers. It is caused by the use of old and improperly stored hops. In some white wines this can be a good thing if it is faint.

Chill Haze A haze formed by protein complexes when the beer is chilled, reducing clarity. Impacts flavor at high levels.

Chocolaty A term most often used to describe rich brown beers such as Porters and Stouts, it describes the flavors and aromas associated with chocolate or dark malts. It is also applicable to certain intense red wines.

Cigar Box Aroma of tobacco and cedar. An especially desirable trait in deep red wines. Rare in ales.

Citrus Aroma and flavor from the citrus family of fruits (grapefruit, orange, lemon). In beer these notes are derived from the hops that were selected (see page 26).

Claret Generic name for a red Bordeaux. Not used in France.

Clean A beer that is light and has no obvious flaws or unwanted aromas and flavors.

Clear A beer or wine with no visible particulate matter.

Clos A French wine term meaning the vineyard is surrounded by a wall. It is of no real importance in this age.

Closed Very little aroma. See Nose.

Clovey Aroma of clove spice. Often associated with German Hefeweissen beers and caused by esters due to the specific yeast strain used.

Cloying A sticky or sickly sweet character that is not balanced. In beer, often associated with too much malt or not enough hops to balance. In wine, too much sugar.

Coarse A term for a beer with a rough texture or mouthfeel. Usually applies to the perception of tannins (especially in red wines), husk flavors or a harsh bitterness in beers.

Coconut Aroma of coconut derived from treatment in American oak in both wines and barrel aged beers.

Coffee Having the roasty aroma and flavor of coffee without being burnt.

Cold Crashing By lowering the temperature of beer after fermentation to near freezing, yeast and other solids aggregate so they are easier to remove.

Cold Hopping Same as dry hopping. See Dry Hopping.

Compact Opposite of Open. Dense flavors and aromas.

Complex Multilayered flavors and aromas. The taste changes over the course of several seconds as you drink it. Not as common in beers as in ales. Expected in any good wine.

Cooked A term where the fruity flavors of the beer seem like they have been cooked, baked or stewed. In wine, it refers to improper storage at a high temperature.

Cooked Vegetables An unfavorable characteristic. Aromas and flavors of cooked cabbage, parsnip, broccoli or celery.

Creamy For beers, this term describes the perception of a smooth, creamy mouthfeel. The perception of creaminess is generally picked up at the sides and back of the throat and through the finish. For wine, this can be an actual taste of cream, and it is especially associated with oaked Chardonnay and some of the best champagnes. Taittinger's *Comtes de Champagne* is arguably the ultimate of all creaminess. In beers, Boddington's *Pub Ale* is a good example achieved by pumping nitrogen gas into the beer.

Crisp A pleasing sense of bitterness in a beer. In wine, this is a mix of relatively high acidity and minerality that is mostly associated with dry white wines.

Crust or Sediment The detritus, generally yeast and protein precipitates, that adhere to the inside (usually bottom) of a aged bottled beer. For wine, see White Diamonds.

Definition A perfect example of the style. See Typicity.

Delicate Relating to the more subtle notes. In ales, the ester aromas from fermentation.

Depth A term used to denote several layers of flavor. A quantitative measure of complexity. See Complex.

Diacetyl or "D" A buttery, butterscotch or buttered popcorn flavor or aroma. Acceptable at very low levels (0.1 ppm or less) but considered a fault at higher levels. One of the vicinal diketones (VDK) is detectable by some people as low as 0.05 parts per million. The source can be yeast metabolism or at higher levels may indicate bacterial contamination (especially when coupled with sourness). See Buttery.

Dimethyl Sulfide (DMS) A sulfur aroma of cooked corn or rancid cooked cabbage in beer or ale. Generally due to insufficient boiling time of the wort.

Dirty A beer with off flavors and aromas that most likely resulted from poor hygiene during the fermentation or packaging process.

DOC *Denominazione di Origine Controllata*. The highest ranking of quality in the Italian classification system.

Douro A wine producing region in Portugal famous for Ports.

Dried Fruit Aroma or flavor of prunes, raisins or dried apricots. Can be applied to ale or wine.

Dry Lacking sweetness. This can be either good or bad, depending on the style.

Dry Hopping The addition of dry hops (typically in pellet form) to beer after the primary fermentation is already complete. Because there is no heat applied, the dry hops are said to only contribute aroma, but technically this is not true. This used to be a technique limited to lagers, but it has gained more widespread use as a general tool of the craft.

Dynamic Generally taken to mean full of flavor for wines and beers, and for beers it also implies strong carbonation.

Earthy Aromas and flavor reminiscent of earth or soil, such as the forest floor and mushrooms. Applies to beer and wine. Quite often used as a subtle warning to the initiated that this beer or wine has a funky aftertaste.

Edgy Sharpness that heightens the flavors on the palate.

Elegant A term applied to describe a beer or champagne that possess finesse with subtle flavors that are in balance.

Enteric A term to describe the vinegar-like sourness common to a young lambic ale.

Estery Aromas of yeast esters from fermentation, often fruity (peach, apple, pear, passion fruit, etc.)

Ethyl Acetate Aromas that are light fruity, pear or solvent-like

Expansive A beer that is considered "big" but still accessible.

Expressive Strong aromas and flavors.

Farmyard A generally more positive term than "Barnyard" used to describe low levels of the earthy and vegetal undertones of some spontaneously fermented beers. It may develop after maturing in the bottle. See Barnyard.

Fat A beer that is full in body and has a sense of viscosity. A beer with too much fat is not balanced and is said to be "flabby". See Full. Seldom used to describe wines.

Figgy Taste or aroma of figs. Uncommon in beer, but common in Gewürztraminer and high-alcohol Zinfandels.

Finish The sense and perception after swallowing.

Finesse A very subjective term used to describe high quality and good balance. Applied to beers and wines alike.

Firm A strong flavor. See Tight .

Flabby Lacking sense of balance, too full or overly thick. The opposite of Tight

Flamboyant In wine, this usually means an excessive amount of fruity sweetness. In ale, it can mean very high alcohol.

Flat In relation to sparkling wines and carbonated beers flat refers to the loss of effervescence. In may also denote a beer that is lacking complexity and finesse.

Flowery Aromas of fresh flowers often from hops or a combination of hops and yeast esters. Belgian ales are mostly characterized this way.

Food Friendly Best consumed with food. This is not a compliment for either beer or wine from a technical standpoint. That's not to say you won't enjoy it, but it doesn't stand up to drinking on its own.

Fresh The perception that the beer has been recently fermented and chilled. What you would expect if you were tasting it right at the brewery.

Fruity The perception of fruit characteristics. This can mean flavors of pineapple, apricot, banana, peach, pear, apple, mango, prickly pear, nectarine, raisins, plum, dates, prunes, figs, blackberry, blueberry, strawberries, and more.

Full or Full Bodied A beer or wine with strong flavor, as opposed to a light lager or a typical Rosé.

Gassy Over carbonated beer from dissolved carbon dioxide (CO_2). Hoegaarden's *Grand Cru* is a perfect example of this, although it is intentional in that case.

Goaty Having the musky aroma of a goat. Rarely good.

Gooseberry This is a berry native to Europe and Africa but relatively unknown in the United States. The flavor is unique, but reminiscent of a cross between green grapes and tart apples. Often associated with Sancerre, an AoP French wine.

Grainy Aromas or flavors of raw grain or cereals, usually a negative descriptor.

Grapefruit Aroma of grapefruit and citrus. May be hop derived.

Grapey Aromas and flavors reminiscent of fresh grapes. Acceptable in Kriek, but generally not a good thing in other beers and ales. In wine it means it lacks alcohol.

Grassy A term used to describe an herbaceous element in a beer ranging from freshly mown lawn grass to hay, alfalfa, straw or open fields.

Gravity (high or low) The original sugar content of a wort before the yeast ferments it into beer

Green Typically used to describe a beer that is not yet finished maturing in its flavor and aroma profile.

Grippy or Gritty Very astringent. In red wine this means excessive tannins.

Gueuze Most Lambic ales include fruits, especially cherries (see Kriek), but Gueuze is the pure Lambic that contains only barley malt, unmalted wheat and hops. It is an acquired taste, being reminiscent of apple cider with vinegar and some herbal notes. It is not a beer in the usual sense.

Harsh Similar to "coarse" but usually used in a more derogatory fashion to denote a beer that is unbalanced in tannins, husky notes, phenols or acidity.

Hazy Having haze, particulates or cloudiness. Not bright

Head This refers to foam on the top of the beer. The foam head should be thick, dense and tight for most beer styles. Some terms for describing a beers head are; persistent, rocky, fluffy, dissipating, lingering, frothy, tight, dense, smooth

Heavy Generally a derogatory term for a beer that is strong in flavor and alcohol. See Robust.

Hefeweizen or Hefeweissen A style of wheat beer. In Germany it is Weissbier, and in Belgium it is Witbier. Traditionally the Belgian type has coriander instead of hops.

Herbaceous An herbal, botanical aroma and flavor. In beer, this is usually from the type and amount of hops used.

Hollow A beer term meaning low in malt or body.

Hoppy Although bitterness is what we mostly associate with hops, the aromas can be flowery, fruity and/or herbal. Hop flavors have a wide range as well which can be astringent, bitter and/or herbal depending a many factor.

Horsey or Horse Blanket An aroma of mustiness with earthy undertones and often a hint of horse sweat. Usually a by-product of *Brettanomyces* yeast and is a character of some beers. Especially in lambic beers.

Hot Judged to have an excessively high ABV.

Husky A flavor of harsh astringent bitterness from grains.

Inky Very dark coloring and high opacity. This is one way to gauge the quality of many red wines. It is also synonymous with German *Schwartzbier*, which translates to "black beer".

Intellectually Satisfying A term coined by legendary wine critic, Robert Parker. Based on the wines that he applies this to, it means an astronomical price.

Intensity The degree of character or strength. Some associated descriptors include complex, hearty, bold, etc.

Jammy Rich in sweet jam-like fruit flavors. Rarely a good thing in beers, but expected in many red wines.

Juicy For wine, this means it still tastes like grapes instead of wine. For Kriek, it means a fresh and sweet taste.

Kriek The most common and popular type of Lambic ale that is made with cherries. See Lambic.

Lace Only in beer and ale. The white foam left on the interior sides of the glass after taking a sip. This is taken as a measure of the quality if it looks like fine lace.

Lagering The slow low-temperature fermentation that is used to produce lager. Note that craft lagers are usually quite different from the commercial types produced in large vats.

Lambic A type of sour ale from Belgium in which barley and wheat are fermented by wild yeast and bacteria native to the region (it is just exposed to the air overnight to begin the fermentation process). Dozens of microorganisms are involved and it cannot be duplicated anywhere else in the world. This is another type of ale good for use in cooking. When it comes to drinking, it is almost unrecognizable as beer. It is completely in a class of its own. See Gueuze.

Lean Wine or beer that is thin, or lacking in complexity. Not quite as thin as the descriptor Watery. See Watery.

Leathery Having an aroma of leather. A positive trait in many intense red wines, and also in German Dunkles Märzen.

Legs The tracks of liquid that cling to the sides of a glass after the contents have been swirled. Often said to be related to the alcohol content of a beer.

Lemony A misleading term that frequently refers to the tangy hoppiness of a beer rather than actual lemons. In wine, it means just what it says—citrus notes of lemon.

Licorice The concentrated flavor in rich sweet beers, often with hints of wood or anise. Especially in Dopplebocks. This is also a subtle flavor note in some intense red wines.

Lightstruck A beer that has had exposure to light causing skunky aroma and flavor.

Linalool Flowery-peach aroma. Derived from the type of hops.

Luscious Similar to voluptuous, but most commonly associated with more fruity flavors, especially for wines.

Meaty A beer that can also be described as brothy or having a Marmite aroma. This is usually from autolysis of yeast cells due to heating too rapidly. A tiny amount in dark strong beers can add depth, but more often it means it's spoiled.

Medicinal Beer with an unpleasant aroma of medicine, paint or disinfectant. This can be the result of improper cleaning of fermentation tanks by amateur brewers.

Melon Aroma of ripe melons. Especially common in Belgian ales, but may be present in any ale. Also in white wines.

Mercaptan Aromas of rotting garlic, dirty drains or an outhouse stench. Uncommon and always a disastrous flaw.

Metallic Aroma or flavor of metal or rust. This can be "Brett Taste" from *Brettanomyces* yeast. See Brett. In white wine it can mean an excessive minerality. See Minerality.

Midpalate A term for the feel and taste when held in the mouth.

Minerality Flavors of slate, rock or mineral water notes. Usually from hard water in the brewing process. For beers, almost exclusively in pale lagers. In wine, this is expected in unoaked Chardonnay and Sauvignon Blanc.

Molasses Aromas and flavors of raw sugar cane, molasses, or black treacle. Also sometimes referred to as Cracker Jack flavor. Molasses notes may occur in sorghum based beers (used for making gluten-free beers). Rare in wine.

Moldy Smells like a damp cellar. Leaf-mold and notes of decay.

Mouthfeel A tasting term used to describe how something literally feels in the mouth. Some commonly applied adjectives are creamy, smooth, silky, opulent, voluptuous, tingly, warming, viscous, oily, coating, thin and watery.

Musky A complex aroma of sweat, sweetness, and earthiness, with light undertones sandalwood. Almost exclusively found in dark beers. A musky wine implies it's spoiled.

Musty Aromas of mold, mildew or decay. A defect. See Moldy.

Nose A term for the aroma or bouquet of any beer or wine.

Nutty Flavors and aromas of various types of nuts except almonds, which would be described as such. See Almondy.

Oaky A beer with a noticeable perception of the effects of oak. This can include the sense of vanilla, butteriness, sweet spice, diacetyl , toasted flavor or woodiness.

Oily A viscous mouthfeel. Can be good or bad in thick ales, depending on your own preferences. Not used for wines.

Open A wine that is fully matured and ready for drinking. Does't apply to beers, other than the bottle being open!

Opulent A rich taste with a pleasing texture and mouthfeel that is well balanced. Applies to beer and wine alike.

Over Carbonated An excessive amount of carbon dioxide dissolved in the beer. This will cause a carbonic bite and or excessive fizziness. See Bite.

Overtones The secondary flavors and aroma of a beer or wine.

Oxidized A negative term describing prolonged exposure to oxygen, giving beers a "wet cardboard" aroma and wines a sour taste. However, in some aged beers that are mildly oxidized, it can bring out a pleasing sherry-like aroma.

Palate A confusing term used for the feel and flavor of a beer or wine in the mouth.

Peppery Aroma and flavor reminiscent of black peppercorns. Seldom encountered in beer, but a common note in Syrah, Zinfandel, and other strong dry red wines.

Perfumy For beer, a generally negative term used to describe an aroma that seems artificial or too floral. This can be a positive attribute in a few types of wine, particularly Rosé.

Phenolic Aroma of plastic or tar. In very faint levels it is acceptable in certain beer styles . It is never acceptable in any wine.

Phenylethanol Floral aroma of a rose-like nature, but not to the point of being perfumy. See Perfumy.

Plonk A derogatory term for cheap mass produced wines.

Plummy The flavor of plum, but not jammy. See Jammy. A component of complexity in red wines and some dark beers.

Port-like Often applied as a criticism of Zinfandels that are very concentrated and high in alcohol. However, enthusiasts of this style will pay high prices for such wines. Martinelli is one vineyard that specializes in these. This term is also applied to some of the most concentrated potent ales.

Powerful A beer or wine with a high level of alcohol but is still balanced.

Proof This is the alcohol (ABV) multiplied by two. This is usually used only for strong liquors rather than beer or wine. Pure grain alcohol (100% ethanol) is 200 proof.

Resinous Aromas of resin, cedar wood, pine, pinewood, sprucy, terpenoid, sap. Can be positive if very faint.

Rindy In beer or ale, an unpleasant bitter aftertaste of hops.

Robust A full bodied beer. Those who do not like this type of beer often use the term Heavy in the pejorative sense.

Round A beer or wine that has a full body but is still balanced.

Sharp A term normally used to describe the acidity of a beer though it can refer to the degree of bitterness derived from a beer's hops. For wine, this usually implies high acidity.

Silky A rich, smooth mouthfeel. Applies to both beer and wine.

Sherry-like Most often used to describe a beer that exhibits sweet sherry-like oxidized aromas.

Skunky Foul odor from improper fermentation. See Lightstruck.

Smokey In wine, this should only be a subtle background note, such as in Pinot Noir. There are strongly smoked German beers called Rauchbier that are produced from smoked malt. *Schlenkerla* is well known among Rauchbiers.

Smooth Soft notes, relatively low alcohol and acidity with no particularly strong characteristic.

Soapy The unpleasant aroma or flavor of soap. See Medicinal.

Soft A beer that is not overly hoppy or overly carbonated. A wine that is fruity and low in alcohol. Rosé wines and medium-sweet wines are usually soft.

Sour A tart flavor, often acidic, sometimes puckering. Appropriate in some beer styles (Belgian browns, Lambic beers, etc). In wine, this is usually due to spoilage.

Spicy Aromas and flavors reminiscent of cloves, coriander, cinnamon and allspice. Tempranillo exhibits this property. In ales, this can be a characteristic of the yeast strain, although spicy notes are also imparted from certain hops, or even adding those actual spices, as in some Belgian ales.

Stale Aromas and flavors of old and oxidized beer, flat and papery, possibly over aged, or over pasteurized.

Stalky A woody, green herbaceous note. Usually unpleasant.

Steely A somewhat acidic and mineral-forward wine. In beer, this often means that it tastes like the can. Japanese Sapporo beer is sometimes cited as an example of this.

Structure A term used to describe the components of a beer or wine's balance. Specifically, how the flavor components harmonize—the collective primary notes of the beer or wine.

Sulfur Aroma of sulfur. Obvious a defect in all wines, but in some beer styles at very low levels it can add to fullness.

Sulfidic Having aromas of strong sulfur, rotten eggs, or natural gas. A serious defect in any wine or beer.

Sulfitic The aroma of a burnt match. A defect in all beer styles, but in some red wines this can be positive in a tiny amount.

Supple A beer or wine that is not overly dynamic.

Sweet For beer, a high level of malt. Also usually lacking in hop balance. Whether this is a defect is a matter of taste. For wine, sweet is the opposite of dry. *Demi-sec* is sweet.

Tannic Aggressive tannins the cause an astringent taste.

Tart Wine or beer with pronounced fruity acidity. This is distinct from a sour vinegar or other foul acidic taste.

Terroir A French wine term for the region where the grapes were grown.

Texture A term for the mouthfeel on the palate.

Thick A beer that is overly full in palate and mouthfeel.

Thin A beer or wine that is lacking body and mouthfeel.

Tight For wine, one that is not sufficiently aged. It may benefit from decanting. For beer, a compliment that it was put together well, properly matured and optimally served.

Toasty For beer and ale, notes from malt character like that of toasted bread. Can also be a sense of the charred or smoky taste from a beer aged in oak wood barrels. For wine, this is taken to mean caramel notes on the finish.

Toffee Sweet flavor of caramel or cooked sugar. For beer, this comes from toasted crystal malt being added. In wine, it is associated with fermentation in new oak barrels.

Trappist Ale Produced by monks in one of only eleven monasteries, six of which are in Belgium. The trappist ales of the world are are Achel, Chimay, Gregorius, La Trappe, Orval, Spencer, Rochefort, Tre Fontane, Westmalle, Westvleteren, and Zundert.

Turbid or Turbidity Cloudiness due to particulate matter. For beer this is usually due to the intentional inclusion of live yeast, especially in Hefeweizens. White wines should be clear. Some red wines, especially very old bottles, can have turbidity for different reasons and it does not necessarily mean there is anything wrong. The solids are usually tannins or tartaric acid. See White Diamonds.

Typicity How well the beer or wine represents the style that it is supposed to be. A lager should taste like a lager, and a cabernet should taste like a cabernet, etc.

Under Attenuated Not a fully fermented beer. Having flavors or components of wort or wortiness. See Worty.

Undertone The more subtle nuances, aromas and flavors.

Upfront Very perceivable characteristics and quality that do not require thought or effort to take notice of.

Vanilla Characteristic aroma of vanilla. Often from oak barrels in either ales or wines.

Vegetal Aromas and flavor reminiscent of vegetation either cooked or raw. In the case of cooked, as in cooked greens or cooked cabbage, parsnip or celery. Usually very bad.

***Vin de Pays* (VdP)** The third level of quality in the French four-tier wind classification system.

***Vin Délimité de Qualité Supérieure* (VDQS)** The second highest level of quality in the four-tier French wine classification system. See AoC.

***Vin de Table* (VdT)** The fourth, or lowest quality standard in the French four-tier wine classification system.

Viscous An ale with excessive heaviness in the body or mouthfeel. See Oily.

Vinous In ale, an aroma, a flavor or texture suggesting wine. In red or white wine, the aroma of grape vines and leaves. Most frequently in the wines of Eastern Europe and Greece.

Voluptuous Full body and rich texture. The opposite of thin.

Warm or Warming In beer, a noticeable but balanced alcohol as opposed to a beer with excessive alcohol that might be described as "hot". A term frequently applied to sherries.

Watery Beyond merely lean, a wine or beer that is watery tastes like it was diluted with water. Coors Light is an example of this, as are many other light beers. Less common in wines. See Lean.

White Diamonds The tartaric acid crystals that may form in the bottle and/or on the cork of good wines. Although aesthetically they may be disagreeable, they are not an indication that there is anything wrong with the wine.

Woody A collective term used to describe the woodsy aroma of a beer that has been treated with oak or other wood. In wine it means a pronounced (or excessive) taste of barrel aging.

Worty Having a taste of wort or unfermented beer, usually disagreeably sweet and lacking complexity.

Yeasty A yeast aroma in beer can be either good or bad, depending on the style and how much of this aroma is present. See Biscuity. No wine should smell of yeast. It means it has spoiled, or (more likely) was made improperly.

Young Wine that has not been aged long enough. Seldom applied to beers. See Green.

Zesty A beer with strong carbonation and some acidity. A wine that has notes of spice, especially bay, black pepper and cloves. Most often Chianti and dry red wines from Spain.

A Few Final Words

A few last minute additions I felt couldn't wait until Volume 4...

POLYPHENOLS

A CONTINUATION OF THE TOPIC FROM PAGE 76

As stated in earlier in this book, there are thousands of polyphenols naturally present in every food we eat—and that's only counting the ones that have been discovered and isolated. One way in which they affect our taste perceptions of foods may be through binding to the enzymes in our saliva that convert chemicals in the food we are chewing to other substances that we actually taste. One important enzyme of this sort is *amylase*, which converts starch to sugar. It has been shown that some polyphenols bind to this enzyme, for example. However, there are dozens of enzymes in human saliva, some of which are still completely unknown in their function! What we do know is that many of these alter the taste of what we are eating, and produce volatilized components that travel up to our nasal cavity where we perceive tastes that are often not in the food before we chew it. Polyphenols are undoubtedly a factor, because we can taste a difference in the concentration and type of polyphenols, but the full scope of this effect is hopelessly complex.

The polyphenol, *quercetin*, is one of the flavonols found in many foods and wines (see page 86). It is now being used widely as an additive in processed turkey meat. This is new technology, though. In the past, manufacturers stuck to manipulating primary flavor components. What they discovered was that the addition of quercetin resulted in taste testers reporting that the meat did not seem to have any artificial flavoring, even when it was loaded with other chemicals. This discovery will undoubtedly be exploited more in the future so that the public is not able to recognize artificially flavored foods. Stick with buying raw ingredients and you'll be fine.

* J. Xiao, et.al. "Interaction of Polyphenols with α-amylase in vitro", *Molecular BioSystems*, **2011**, (7) pg. 1883-1890

BEEF AND VEAL STOCK IN A PRESSURE COOKER

A CONTINUATION OF THE TOPIC FROM PAGE 94

It might seem that the way to get around the limitation of large bones taking too long to dissolve when attempting to make stock with a pressure cooker is simply to add granulated gelatine with some meat and aromatic vegetables. It doesn't work that way, though. That concoction cannot be called stock. What you get is a dark broth thickened with gelatine. Plus, that gelatine will break down when you try to use it in cooking sauces. Not all gelatine is created equally. True stock also contains a panoply of polyphenols and other secondary flavor molecules that were released slowly at a lower temperature. Stock prepared in a pressure cooker with gelatine has a harsh taste to it, for lack of a better term. Stock prepared by slow simmering for many hours has a very soft taste with a lot of unctuous mouth feel that is not just the gelatine, but the physiological effects of those polyphenols, as previously explained. Powdered gelatine and pressure cookers have been around a long time. If such a shortcut actually worked everyone would use it.

RELIABILITY OF THE INFORMATION IN THIS BOOK

I never offer advice about something I'm unsure about myself. The information presented here is, to the best of my knowledge correct in every detail. Aside from my university education in chemistry and decades of experience cooking, the topics in this book have been researched at length from credible scientific sources. For example, here are some of the sources drawn on for the chapter on onions...

Journal of American Horticultural Science, **1995**, *120* (6), pp 1075-1081
American Journal of Clinical Nutrition, **2000**, *72* (6), pp 1424-1435
HortScience, **2002**, *37* (3), pp 567-570
Journal of Agriculture and Food Chemistry, **2004**, *52* (22), pp 2797-2802
Journal of Agriculture and Food Chemistry, **2004**, *52* (22), pp 5383-5390
Journal of Agriculture and Food Chemistry, **2004**, *52* (22), pp 6787-6793
Journal of Agriculture and Food Chemistry, **2007**, *55* (25), pp 10067–10080
Journal of Agriculture and Food Chemistry, **2010**, *58* (4), pp 2323-2330
Annals of Botany, **2012**, *109* (4), pp 819-831
Journal of Culinary Science and Technology, **2016**, *14* (1), pp 1-12
Culinary Herbs (Second Edition), Ernest Small, **2006** NRC Research Press
Onions and Allied Crops, Volume 3, Brewster and Rabinowitch, **1989** CRC Press

However, no one is infallible. Please let me know if you find any mistakes or omissions and I will report them as such in Volume 4.

OLIVE OIL PRODUCTION

A CONTINUATION OF THE TOPIC FROM PAGE 63

The diagram below explains the steps involved in producing extra-virgin olive oil.

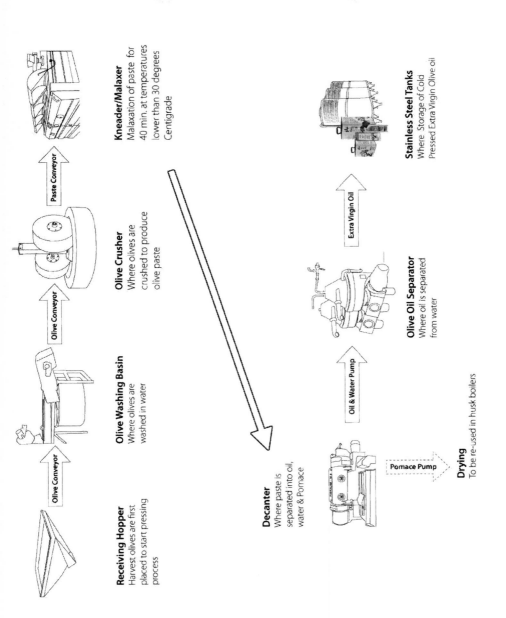

Kneader/Malaxer
Malaxation of paste for 40 min. at temperatures lower than 30 degrees Centigrade

Olive Crusher
Where olives are crushed to produce olive paste

Olive Washing Basin
Where olives are washed in water

Receiving Hopper
Harvest olives are first placed to start pressing process

Paste Conveyor

Olive Conveyor

Olive Conveyor

Stainless Steel Tanks
Where Storage of Cold Pressed Extra Virgin Olive oil

Extra Virgin Oil

Olive Oil Separator
Where oil is separated from water

Oil & Water Pump

Decanter
Where paste is separated into oil, water & Pomace

Pomace Pump

Drying
To be re-used in husk boilers

WHAT DOES FAN ASSIST IN OVENS DO?
A QUESTION BY A VIEWER

There are actually two types of fan assisted ovens. One has a heating element behind the fan, and it blows very hot air around inside. The other type, which is more common, has heating elements in the top and bottom of the oven and a fan that circulates the air, as well as venting it out the top back of the stove. Fan assist is more even, but also has a significant drying effect that aids in crisping foods. If you are baking cakes, then the benefit of fan assist will depend on the recipe. If the batter is very moist, then it can be a huge help. If not, it can cause problems. If you are shopping for a new stove, be sure to get one with fan assist, but also be sure that there are controls that enable you to turn the fan on and off. Usually there is a rotary dial with various modes that set top and/or bottom heating elements and the fan on or off. Don't get a stove that runs the fan at all times.

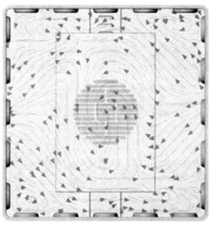

Air-flow with forced air circulation by fan

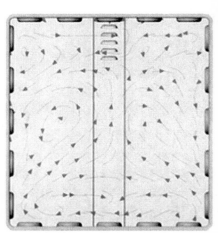

Air flow with natural convection

HEMP FLOUR

Hemp flour is available in health food stores and online. While marijuana is high in THC (the psychoactive component), hemp is very low in this. Although it is still illegal to grow hemp in the United States, it is perfectly legal to import, so hemp flour and other hemp products are all imports. It is available in Russia, but not common.

ALEPPO PEPPER AND RUSSIAN CUSTOMS

This is actually a type of dried chili and not a type of peppercorn. It is most frequently seen as flakes, but also sometimes as a ground powder. Up until fairly recently it was only used in Middle Eastern recipes, but western chefs looking for more exotic spices discovered it, and now you see it being used in untraditional ways. In particular I have seen this called for in Italian recipes lately in lieu of ordinary crushed red pepper flakes. The only reason that I never use it is because it is not available in Russia, and the original theme of my channel is restricted to things that you can cook in Russia with ingredients that are available (even though some of them do take some searching to find). In case you are thinking that this problem could be solved by buying online... no, it can't. Russian customs doesn't allow food products through—and when it is a substance that they are not familiar with, the first assumption is that it is some kind of illegal drug. The problem is that very, very few Russians have any interest in gourmet cooking, and the type of person who's a customs agent is someone who earns a low wage and eats simple traditional Russian foods. You can get ingredients like this in through the airport in your luggage much easier, but there's always the risk of running into serious problems.

Several years ago I complained to one of my regular customers at a restaurant where I was the head chef, about baking powder being unavailable in Russia. Baking soda is everywhere, but double-acting baking powder is nearly impossible to find. It's pretty easy to become friends with any native English speaker in Russia, because we are a small group. He also happened to be the CEO of a large corporation in Europe and he flew into Russia quite often. A couple of months later he brought me a dozen bottles of baking powder in his baggage via the airport. Only his baggage had been banged up some, and a bottle or two split open. He regaled me with the tale of how he was in the airport with white powder visibly leaking out of a suitcase. He was pulled off to the side and put in a locked cell while they investigated what it was. Luckily, he had the connections and power to get himself out of that mess. He made a phone call and a few minutes later the customs agents cleared him through with smiles and apologies. Lucky him. If you don't have such powers, never try to bring anything into Russia that might look like drugs.

DRIED CHIPOTLE CHILIES

The only dried chili peppers that exist in this part of Russia are crushed red pepper flakes, which are cayenne (although sometimes manufacturers mix in other chilies to maintain consistent heat levels from batch to batch). Very few Russians like anything spicy, so the selection of fresh chili peppers is extremely limited and requires a great deal of searching. After a two hour trek by bus, train and a very long trolley ride, I managed to locate the holy grail—canned chipotle chilies in adobo sauce from Mexico. They were on the same shelf as the black truffles and priced about the same per gram. I needed dried chipotles in some of my old pre-Russia recipes and drying is a good way of preserving such things. So I rinsed some off and dried them in the oven using the same method I described in Volume 1. The result was actually better than I had expected. Although they are not as smoky as the commercially sold chipotles, they are more flavorful and brighter tasting, just like other chilies you dry yourself. The lack of smoke can be cured with the addition of a little liquid smoke to whatever it is you are making. This is an experiment you should definitely try.

GMO CONTROVERSY

This is another example of the public largely succumbing to fear mongering. While there are *some* valid health concerns, it should also be noted that as the planet will soon have 8 billion humans (predicted date 2024) and most of us would be dead from serious malnutrition already if not for the technology that enables plants to be resistant to insects, diseases, drought and offer important advantages in harvesting. You should know that more than 100 Nobel Laureates including cell biologists and agricultural scientists recently signed a petition against Greenpeace asking them to stop making irresponsible blanket statements against GMO's. The fact is that many GMO products are safer and healthier than the non-GMO variety! So when you are in a store and see a product advertised as not having any GMO, don't jump to the conclusion that it is better. It may very well be worse for both you and the planet. The bottom line is that if you aren't an expert in the field, you probably don't have the necessary information to make a truly informed decision. A "No GMO" sticker is often just marketing.

*For several years I was a regular columnist for the Los Angeles Times in
their Food & Wine section. One of the last articles I ever wrote was on
sous vide cooking. This is a topic that I'm frequently asked about in
comments on YouTube. Here is a reproduction of that column.*

SOUS VIDE

AN EDITORIAL

At first the biggest fans of sous vide cooking were restaurant owners.
Here was a method that could be used to produce good food by unskilled
labor. Stick food in bag. Seal bag. Put bag in water bath. If you
accidentally leave it in for an extra hour, it's probably still fine. No need
for trained chefs with this sort of technology, right? The dream of every
shrewd business owner is to cut costs by replacing the highly paid skilled
workers with gadgets. Gadgets pay for themselves in saved wages.

Then some famous chefs began promoting this technique. After all, no
one can rightfully argue that sous vide does not have its merits. But so
does a hammer if you are trying to drive a nail. It's just about the best tool
you could find, in fact. Way better than a wrench or a rock. Except not
every task in building involves pounding nails. Using a hammer to build a
chimney is absurd. Some people try to apply sous vide to everything. I'm
frequently asked if sous vide can be used to prepare recipes that involve
frying and roasting? It's like asking if a socket wrench can be used to
install plate glass windows? No. Put the toy away and learn to cook.

One of the most poignant arguments I have heard against sous vide is
that it sterilizes the act of cooking itself. You no longer experience the
aromas and sounds of foods being cooked. You are left out of the process
as bags silently swirl around in their little vacuum filled universe on their
own until a bell rings to signal you to salivate. I don't mean to say that
sous vide is not a fabulous tool when properly wielded. My complaint is
that it is being marketed to people who can't even boil pasta—as if it's
going to transform them into Michelin star chefs through the magic of
technology—while in fact it is doing the *exact opposite!* Rather than
learning to cook, they pop food into a bag and go back to watching TV.

Egg

Index

A

B

C

D

E

F

G

H

I

J

K

V

W

X Y Z

CPSIA information can be obtained at www.ICGtesting.com
Printed in the USA
LVOW11s0240170816

500723LV00001B/71/P